定制办公空间
Custom Office Space

ID Book 图书工作室 编

华中科技大学出版社
http://www.hustp.com
中国·武汉

序言 | Preface

办公设计新思维

在现代社会中，人们除了重视自己的生活质量，也更多关注起自己的工作环境与氛围。在我们的生活中，至少有一半时间是待在工作室里。现代生活快节奏的脚步，来自于生活中各个方面的压力，使我们总是绷紧了神经。这让我们更加渴望有一个舒适的办公环境及轻松的工作氛围。快乐办公，高效工作，成为当代办公空间设计的新内涵。

无论是办公室，还是工厂、实验室，大到大众公共场所，小到私人私密空间，任何类型的空间设计都需要我们根据特定的需要，协调整个空间布局、材料预算、软装陈设等的种种矛盾，将其融合为一体，使整个空间的设计理念更加人性化，形象更加美观，主题更加鲜明，功能更加实用。

一个优秀的办公空间的设计，能从侧面反映出一个企业的文化内涵。它能提高员工的工作效率，并协调团队之间的合作氛围，能使人以更加轻松的状态创造出更加高效的成果。它能使人的潜能得到激发，让人与人之间的行为方式相互影响、相互校正，最终成为一个优秀的员工。一个出色的办公空间设计，在权衡好以上需求的前提下，才能让工作其中的人们充满归宿感。

广州道胜装饰设计有限公司
2013年10月21日

New Thinking on Office Design

In modern society, apart from their own life quality, people pay more and more attention to their work environment and work atmosphere. In our life, half of our time is spent in the work place. With the fast pace of modern life, pressure from all aspects of life always makes us tighten the nerves. And that makes us yearn more for a comfortable office environment and relaxing work atmosphere. Happy work space and high-efficient work become the new connotations for modern office space design.

Be it office, factory or laboratory, as big as public space, as small as private personal space, each kind of office space would always require us to coordinate all contradictions on space layout, materials budget, soft decorations, and so on, according to the specific requirements and integrate them as a whole, making the whole space design concept more humanized, the image much better, the theme more distinct and the functions more practical.

An excellent office space design would reflect the corporate cultural connotations from some aspects. It can uplift the work efficiency of the staff and produce some cooperative atmosphere among the team members, making people create some results with higher efficiency in more relaxing circumstances. It can make people make full use of their potentials and produce mutual influence and adjustment between people's behavioral patterns, finally creating some excellent staff. Upon balancing well all the above requirements, can an outstanding office space design make people work inside get some sense of belonging.

Guangzhou Daosheng Design Co., Ltd.
Oct. 21st, 2013

目录 | Contents

006	Tender Luxury Tender Luxury	084	万科钻石广场Loft A4办公样板房 Vanke Diamond Square Loft A4 Office Model House
014	保利商业地产办公室 Office for Poly Commercial Real Estate	090	新创广告办公室 Office for Xinchuang Advertising
020	环宇建筑办公室设计 Office Design for Huanyu Construction	096	东方国际•创冠集团香港总部 Oriental International•C&G Group Hong Kong Headquarters
024	LAVIE公社销售办公室 Sales Office for LAVIE Commune	108	办公空间的光影魔术 Light and Shadow Magic of Office Space
034	道胜设计新办公室 New Office for Daosheng Design	114	鸿隆世纪广场A+B户型办公样板间 Hong Long Century Plaza, A+B Office Model Room
040	福州胜道设计公司办公室 Office for Fuzhou Shengdao Design Company	120	Angular Momentum Angular Momentum
046	集叁设计工作室 Jisan Design Studio	132	前线共和广告公司办公室 Office for Frontline Republican Advertising Office
050	开山设计顾问办公室 Office for KSEN Design Consultants	138	树林里的办公室 Office in the Forest
054	易•空间办公设计 Yi • Space Office	144	唯知唯美 Purely Aesthetic Space
060	太空利器 Space Edge Tool	150	道和设计机构办公室 Office for Daohe Design Institution
066	中企绿色总部•广佛基地办公室 ZhongQi green headquarters • guangfo base office	156	世纪嘉美办公室 Shiji Jiamei Office
074	卓越世纪中心办公式样板房 Excellence Century Center Office Style Sample House	162	保发大厦劳伦斯珠宝写字楼 Baofa Building Lorenzo Jewelry Office Building
078	HED+OFFICE HED+OFFICE	168	石油化工交易所办公空间 Office Space for Petrochemical Exchange

176	波龙办公室 Bolon Office	250	智威汤逊北京公司(JWT) J.Walter Thompson Company in Beijing
182	设计师的办公室 Designer's Office	268	科大永合投资有限公司 Keda Yonghe Investment Co., Ltd.
188	万科润园办公空间 Office Space for Vanke Runyuan	274	天安国际大厦商务楼 Office Building for Tian'an International Tower
194	诗意办公 Poetic Office	282	凯发贸易办公空间 Office Space for Kaifa Trade Co.
198	BEYOND BUY INC I BEYOND BUY INC I	286	凯明迪律师事务所办公室设计 Office Design for Chiomenti Studio Legale
204	东莞虎门实业集团办公室 Dongguan Humen Industrial Group Office	292	IEA总部办公室 Office for IEA Headquarters
210	深圳粤华集团办公室 Shenzhen Yuehua Group Office	298	杨大明办公室 Yang Daming's Office
216	树德办公总部 Shude Headquarters Office	304	云裳•婚纱工作室 Yunshang•Wedding Dress Studio
222	捷致办公室 Jiezhi Office	312	黑白的奏章•力宝天马大厦 Musical Chapter of Black and White•Lippo ianma Building
226	禾大办公区设计 CRODA OFFICE	318	香港华锋实业E路航办公室设计 HongKong Huafeng Industrial Co., Ltd. Office Design
230	Paga Todo办公区设计 Paga Todo Office		
236	墨西哥城某办公空间 Corpovo Ifahto		
244	BEYOND BUY INC II BEYOND BUY INC II		

Tender Luxury

Tender Luxury

设计单位：上海牧桓建筑 + 灯光设计顾问
设 计 师：赵牧桓
参与设计：王颖建、胡昕岳
项目地点：上海市
项目面积：700 m²
主要材料：橡木、黄洞石、玻璃、铁件、白杨木
摄 影 师：周宇贤

Design Company: MoHen Design International
Designer: Hank M. Chao
Associate Designers: Wang Yingjian, Hu Xinyue
Project Location: Shanghai
Project Area: 700 m²
Major Materials: Oak, Yellow Travertine, Glass, Iron, Poplar
Photographer: Zhou Yuxian

奢华的形式大多集中在复杂的表象上面,有点像西方洛可可式的在视觉上眼花缭乱的组合方法,但这并不是表达奢华的唯一手法。

整体色彩呈现出沉稳内敛的基调,呼应在概念上刻意回避的铺张感,但透过材质表象却有一种更深沉的华丽感。项目所释放出来的较为宽阔的空间体量表现出的是另一种概念上的奢华感,这是一种富有张力的视觉感,也是作为设计者想透过本案所传达的一种低调的设计理念。

座位区中,设计师利用从顶棚处垂下来的不到底的书柜作为分

隔，在视觉上有种穿透感。金属帘也运用在此，这是一种软性区隔空间的方式。另外设计师以雕刻的手法设计了吧台的造型，与周围的直线条做了对比，背景是白杨木的树林，光与树影相互穿插，若隐若现。不经意地洒在地面上，增加了诗意感。通往贵宾包厢的走道也同样地将水波投射在地面上，这种动态的方式使人们进入包厢的过程有了趣味，也弥补了在设计上无法与人互动的缺陷。包厢里设置的壁炉增加了视觉上的温馨感。

本案坐落在上海浦东区的黄埔江边，是一个新发展的地块，周围的景观配套有着良好的规划。入口处的设计刻意强调景深，并用水池隔开了通往后方的洗手间。在通往洗手间的墙立面上以水波纹的动态投影投射流动水的缥缈质感，使得室内与室外的黄埔江有了视觉上的联动关系。

Luxury is mostly displayed on the complicated surface, kind of like the western Rococo style group approach dazzling the eyes. But that is not the only approach to manifest luxury.

This project is located along Huangpu River of Shanghai's Pudong District, as a newly developed area, possessing a wonderful planning for the surrounding landscape. The design at the entrance stresses on the depth of views, separated from the back wash room with a water pool. On the wall leading to the wash room, the dynamic water waves convey the misty texture of the water, producing some visual linkage relationship between interior space and the exterior Huangpu River.

For the seating area, the designer makes use of the bookcase from the ceiling as a separation, with visual transparency feel. The metal curtain is also applied here, as a soft approach in separating space. Other than that, the designer makes use of carving approach to design the format of the bar counter, forming contrast with the surrounding straight lines, with as background poplar forest, light and tree shadows interlacing with each other, partly hidden and partly visible. The light and shadows fall on the ground without people noticing it, creating some poetic feel. The corridor leading to the VIP box rooms also has water waves reflected on the ground, with this dynamic method producing some interests for people's walking process into the box rooms, while complementing the flaw

of lacking interaction with people. The furnace in the box rooms adds some warm visual sensations.

The whole color displays some sedate and profound tones, corresponding with the avoided extravagant feel in the concept, with the materials possessing some more intense magnificent feel. The comparatively broad space scale of the project displays some luxury on some other concept. This is some visual feel full of tension and is also some low-key design concept that the designer wants to convey through this project.

保利商业地产办公室
Office for Poly Commercial Real Estate

设计公司：广州道胜装饰设计有限公司
设 计 师：何永明
项目地点：广东省广州市
项目面积：610 m²
摄 影 师：彭宇宪

Design Company: Guangzhou Daosheng Decorative Design Co., Ltd.
Designer: He Yongming
Project Location: Guangzhou in Guangdong Province
Project Area: 610 m²
Photographer: Peng Yuxian

办公空间的设计，要营造出符合企业精神，并且舒适、健康、人性化的办公环境，展示出先进的工作理念和进取开放的企业文化，同时也强调员工之间的互助与团队精神，从而增加员工之间的凝聚力和归属感。

设计师以一长轴线串联所有空间，作为空间的主要动线。空间布局上，在大门入口设置了前厅接待区，接着是会议室一，再以中间过道为中心，左边布置办公区、会议室二，右边为会议室三、副经理室、水吧休闲区、财务室和总经理室。

员工工作台之间以中低尺度的书柜作为分隔，在保留了工作空间独立性的同时，也方便了员工之间的交流。这种开放的设计维持了视觉的穿透性，让空间更加流畅，同时也呈现出私密与开放并存的工作理念。

整体色调采用暖灰色，这种色调能够使员工的心情沉静下来，提高工作效率。有一位建筑大师曾说过："当绿化、光与人的意念从原先的自然中抽象出来，它们就趋向天堂。"所以我们尽量保留原有的窗户，引入大量自然光，放置大量的绿色植物，将大自然的感觉引入室内空间，使办公环境更具活力。

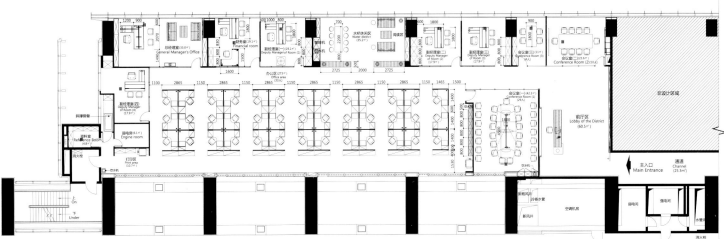

Custom Office Space   015

The design of office space should create comfortable, healthy and human office environment in accordance with the corporate spirits, presenting advanced work concepts and pioneering corporate culture, while strengthening the mutual assistance among staff and the team spirits, adding to the cohesive power and sense of belonging of the staff.

The designer connects all the spaces with a long axis, as the main moving line of the whole space. As for the space layout, at the door entrance is the parlor reception area, followed by conference room 1. With the central corridor in the center, on the left side is office area and conference room 2, and on the right side is conference room 3, Deputy Manager's office, bar lounge area, financial office and General Manager's office.

The staff working tables have medium to low scale bookcases as the partition, while maintaining the independence of the work space, providing convenience for the communication among the staff. Such kind of open design continues the transparency of the space, making the space more fluent, presenting the work concept of the coexistence of privacy and open style.

The whole color applies warm gray color, which can quiet down the moods of the staff, while uplifting the work efficiency. One architectural master once said that "When greening, light and people's idea are extracted from the original nature, they are orienting towards the paradise." Thus we try the best to maintain the original windows, introduce quite a lot of natural light, set a quantity of green plants, introducing the feel of the grand nature inside the interior space, entrusting the office space with more vitality.

环宇建筑办公室设计
Office Design for Huanyu Construction

设计单位：名宿设计
设 计 师：林德华
项目地点：福建省福州市
项目面积：340 m²
主要材料：大理石、地板、铝材、草皮
摄 影 师：周跃东

Design Company: Mingsu Design
Designer: Lin Dehua
Project Location: Fuzhou in Fujian Province
Project Area: 340 m²
Major Materials: marble, floor, aluminum, turf
Photographer: Zhou Yuedong

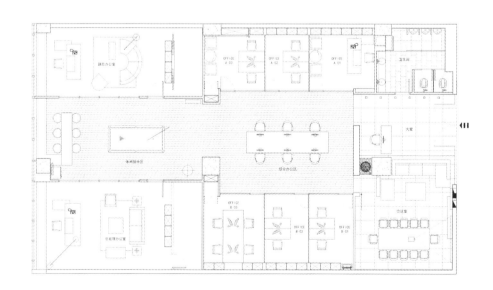

本项目位于一个20世纪中期的建筑中,设计师在保持原建筑工业美感的同时,使用了天然的材质赋予它新的生命。大面积的栽培草平衡了空气中的湿度和温度,天然木纹带来了质朴感,休闲空间的设置使工作氛围更为活泼。这种简洁、环保、轻松的环境给日常的工作带来了积极有效的动力。

公司的前台区域在材料的运用上,以绿色环保为前提,没有过度地装饰墙面,主要以色彩来突显空间的张力。大面积的栽培草广泛应用于外围的墙面和天花,不仅平衡了空气中的湿度和温度,也具有很强的设计理念与创意感。天然木纹地板有一种天然的质朴感,木纹地板上的射灯在栽培草墙面上形成了如阳光似的光圈,设计师通过天与地的相互呼应达到空间的完整与统一。从大门往里看,映入眼前的是一片森林的景象。

办公区呈开放式,不加修饰的钢筋混凝土建筑框架毫无保留地得到展现,而空间布局则结合其建筑框架慢慢铺开。由于整体的层高较低,设计师在分割区域时选用了大面积的玻璃和镜面材料,横向扩展空间以减少压迫感,看似封闭但视线相通的各个办公区形成一个隔而不断、分而不离的办公环境。内部也零星地铺设栽培草墙面,在灯具的选择上使用了环保的LED灯,红色的光线与绿色栽培草在色彩碰撞中浮现出隐约的神秘感。

This project is located in a building dated to the mid of 20th century. While maintaining the industrial aesthetic beauty of the original building, the designer makes use of

Custom Office Space

some natural materials to entrust it with some new life. The large area of turf balances the humidity and temperature of the air, natural wood grain brings some primitive feel, while the leisure space setting makes the work atmosphere dynamic and active. This concise, environmental and relaxing environment brings some active and efficient motive for the daily work.

On materials application of the corporation's reception area, it has green and environment as the premise, with no exaggerated decorations on the wall, mainly using colors to highlight the tension of the space. The large area of turf is also widely applied on the exterior wall and ceiling, not only balancing the humidity and temperature in the air, but also possessing some intensive design concept and innovative feel. The natural wood grain floor possesses some natural primitive feel. The spotlights on the wood grain floors produce some sunshine-like halo on the wall of grass. The designer achieves completeness and integrity of the space through echoing of land and heaven. When you observe from the door, you can find some woods-like sceneries.

The office area is in open style, the armored concrete building framework with no decorations are displayed to the full and the space layout is gradually presented combined with the building framework. Due to the whole low floor height, in the division area, the designer selects some large area glass and mirror materials, releasing pressure while horizontally expanding the space. Seemingly enclosed but visually connected office spaces form some continuous and divided office environment with partitions and separations. Inside the space, there is also turf wall, with environmental LED lights on the lighting accessories selection. Within the color impact, the red lights and the green grass produce some subtle mysterious feel.

LAVIE 公社销售办公室
Sales Office for LAVIE Commune

设计单位：重庆尚壹扬设计有限公司
设 计 师：谢柯、支鸿鑫
项目地点：山东省青岛市

Design Company: Chongqing Shangyiyang Decoration Co., Ltd.
Designers: Xie Ke, Zhi Hongxin
Project Location: Qingdao in Sihandong Province

本案是一个典型的现代主义与古典主义相融合的设计作品，视觉上呈现出简约的气质，感知上给人以新古典的内在精神。虽是两种不同风格的混合，但却结合得恰到好处，是一处极富感染力的销售办公空间，表达了以样板房销售为主的另类办公体验。

在这个两层的办公空间里，设计师在功能规划上充分考虑人性化办公空间设计所应注重的细节，合理安排功能分区，按照人机工程学的原理，合理选择办公家具，使办公人员更为舒适自在，使生活与工作形成一种良性的互动。

在材质选择上，钢结构楼梯、实木吊顶和素色砖墙材质的自然肌理得到最大限度的呈现，巧妙地塑造出空间的质朴本质。更为珍贵的是，设计师将历史文化元素贯穿其中，使项目本身具有了独特的内涵。

Custom Office Space 025

This project is a classical design work of the integration of modernism and classical design, demonstrating some concise temperament visually, while with some inherent neo-classical spirits. Although this is the combination of two different styles, it attains perfect bond. This sales office space of rich infection conveys some different office experiences centered on show flat marketing.

Within this two floor office space, on functional planning, the designer fully considers the details that human office space design should focus on, properly distributes the functional areas and selects office furniture according to the principle of ergonomics, making the staff feel more relaxed and casual and performing some benign interaction between life and work.

In materials selection, steel structure staircase, solid wood ceiling and the natural texture of plain color brick wall display utmost presentation, ingeniously producing the primitive nature of the space. What is more precious is that, the historical and cultural elements run through the whole space, entrusting the project itself with spectacular connotations.

道胜设计新办公室
New Office for Daosheng Design

设计单位：广州道胜装饰设计有限公司
设 计 师：何永明
项目地点：广东省广州市
项目面积：600 m²
主要材料：直纹白大理石、黑白根大理石、地胶、黑镜、壁纸
摄 影 师：彭宇宪

Design Company: Guangzhou Daosheng Design Co., Ltd.
Designer: He Yongming
Project Location: Guangzhou in Guangdong Province
Project Area: 600 m²
Major Materials: straight grain white marble, black marquina, badminton flooring, black mirror, wallpaper
Photographer: Peng Yuxian

Custom Office Space

整个办公空间的设计简洁、明快,设计师以功能为主导、艺术品为点缀,在空间中以黑、白、灰为主色调,营造出高档、时尚的空间。同时也诠释了"以人为本、人居合一"的设计理念。

整个办公空间平面呈长方形,以会议室、洽谈区为中心,分为左、右两部分。左边为设计区,右边为行政区。前台的接待区以石材与镜面为材料,体现出简约设计的分量感。简洁大方的公司LOGO点名主题,入口处的两只小猪为原本沉稳的空间气氛增添了创意性及趣味性。

会议室具有图纸展示、多媒体展示等功能,摒弃了办公椅的传统造型,而是选用了原木的中式椅,为严肃的会议室增添了亲切感。墙面上百鸟归巢的立体画,寓意公司团队凝聚着无穷的力量。

开敞式的设计区,为了满足个人的需求,每个位置都配备了可移动的办公椅及储物柜,在满足实用功能的基础上,用原木设计隔断式的书柜,增强了空间的创意感,也很好地利用了资源。

通往行政区的走廊,用精美的艺术品与寓意深远的挂画打破了狭窄的视觉感受,很好地实现了人与空间的对话。

总经理办公室是个多功能的组合空间,办公室与茶室不但满足了办公功能,还具有会客、休闲功能,选用的家具和落地灯营造出轻松、雅致的办公环境,书法作品与粘贴于门上的枯树叶,彰显出设计的独到品位。

The design of the whole office space is concise and brisk. Guided by functions, ornamented with artistic objects, the designer creates some high-level and fashionable space with black, white and gray as the tone colors, while interpreting the design concept of human-oriented principle and integration of human and residence. The whole office space plane is rectangular, centering on conference room and negotiation area, and divided into east and west two parts. To the left is the design zone and the right administration zone. The reception area has stone and mirror surface as the materials, representing the scale of concise design. Concise and grand corporate LOGO sets the theme and two little pigs decoration at the entrance adds some creation and interests for the original sedate space atmosphere.

The conference rooms have drawing paper presentation and multimedia presentation and other functions, forbidding the traditional formats of office chairs, but selecting log wood Chinese chairs, producing some intimate feel for the solemn conference rooms. The stereograph of homing birds on the wall implies the

immense power of the corporate team.

In order to meet with the requirements of every individual, in the open style design area, every location is accompanied with movable office chairs and lockers. On the basis of practical functions, the log wood is designed into partition style bookcase, strengthening the creative feel of the space, while making good use of the resources.

For the corridor leading to the administration area, the exquisite artistic objects and painting of immense implications break through the narrow visual sensations, well achieving the dialogue of people and space.

The office for the general manager is a multi-functional grouping space. The office and the tea room not only meet with office functions, but also boast meeting and leisure functions, with furniture and floor lamps creating some relaxing and elegant office environment and the calligraphy and writing and dead leaves pasted on the door demonstrate the peculiar taste of the design.

福州胜道设计公司办公室
Office for Fuzhou Shengdao Design Company

主创设计师：叶强

参与设计师：简华辉、陈浩天

项目地点：福建省福州市

项目面积：150 m²

摄 影 师：李玲玉

主要材料：木皮免漆板、白色成品烤漆、黑色金属烤漆、进口艺术肌理涂料

Leading Designer: Ye Qiang
Associate Designers: Jian Huahui, Chen Haotian
Project Location: Fuzhou in Fujian Province
Project Area: 150 m²
Photographer: Li Lingyu
Major Materials: wood plate, white lacquer paint, black metal stoving varnish, imported artistic texture coating

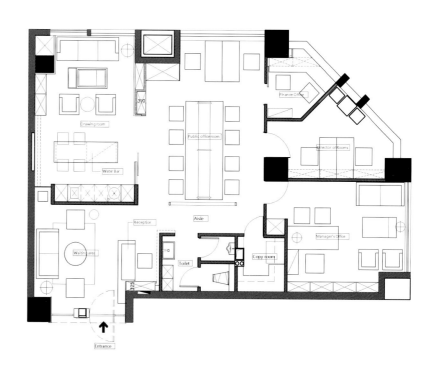

做办公室的设计首先要了解企业类型和企业文化,了解了企业内部的机构设置及其相互联系,才能确定各部门的所需面积,并规划好人流线路,才能设计出反映该企业风格与特征的办公空间,使设计具有个性与生命。

办公室装修要营造安静、平和与整洁的环境。秩序感是办公室装修设计的一个基本要素,这涉及平面布置的规整性,以及家具样式与色彩的统一、隔断高低尺寸与材料的统一等。办公室隔断的设置要尽可能不使个人空间受到干扰,提高员工工作的注意力。另外,应尽量利用简洁的建筑手法,避免采用过多的造型、繁琐的细部装饰、过浓的色彩点缀。在设置灯具、空调和选择办公家具时,应充分考虑其适用性和舒适性,还应在装修设计中融入环保的观念。

本案是一个位于福州城市中心的设计公司办公室,每个办公空间都像被量身定制了一样,既具有开放性,又确保了隐私。原木的材料和灯光的配置都使得这个空间充满了清新、时尚、舒适的气息,这样的工作环境既可提高工作效率,又能给员工提供一个舒心的办公空间。

Office design should first understand the corporate style and the corporate culture, understand the corporate inner institution setting and the relationships among each other, thus the required area for each department can be defined, streamline can be planned, and office space reflecting the corporate style and features can be created, entrusting design with personality and life.

Office decoration should produce a quiet, peaceful and tidy environment. Order is a fundamental element in office decoration and design, which involves the order of plane setting, integrity of furniture format and colors, high and low scale of partition and the consistence of materials. The setting of office partition should try the best to avoid personal space to get disturbed, thus strengthening the attentiveness of the staff. Other than that, the design should maximize the application of concise architectural approaches, avoiding too many formats, much too complicated detail decoration or overrich color decorations.

The setting of lights, air-conditioners and office furniture should fully take into considerations the practicality and comfort, while integrated with environmental concepts in the decorative design. This is an office for a design firm in the urban center of Fuzhou, with every office space like custom-made, being open while preserving the privacy. Log wood materials and lights allow the space to be filled with fresh, fashionable, and comfortable atmosphere, which can not only promote work efficiency, but also provide a cozy space for the staff.

集叁设计工作室　　Jisan Design Studio

设 计 师：邓明发
项目地点：无锡省无锡市
项目面积：280 m²
主要材料：中国黑烧大理石、水曲柳染色、玻璃

Designer: Deng Mingfa
Project Location: Wuxiin Jiangsu Province
Project Area: 280 m²
Major Materials: China black marble, dyed ashtree wood, glass

原本规则的空间被建筑本身中几根零碎的柱子分成九格，设计师就此根据建筑本身的特点对空间做了改造，把不需要的墙体打开，进行区位分隔，将之规划出会客室、接待室、洽谈室、设计室、楼梯间等。

二层的办公空间利用水曲柳自身温馨的暖色调来传达在此工作的人们的友善与团结。在现代的办公空间内利用中国黑大理石烧毛处理来中和白色调的轻浮感，整体空间给人以现代简洁的感觉，但是也透露出几分淳朴的气息。

在规划功能空间的同时，又利用玻璃隔断和白色方管来处理孤立的柱子。白色的方管衔接一层与二层的空间，又将纵向的线条引至二层。楼梯间错落的吊灯好像将要打破楼梯下方平静的水面，油画中的女童拿着荷叶，仿佛也怕身体被打湿。

Custom Office Space 047

The original orderly space is divided into 9 grids with several scattered pillars in the building. Thus the designer carries out space regeneration based on the original features of the building, opening the unnecessary walls, setting zone separation and planning meeting room, reception room, negotiation room, design studio, staircase, etc.

While planning functional space, the designer also makes use of glass separations and white square tubes to deal with the single pillars. The white square tubes connect the first floor with the second floor, while leading the vertical lines to the second floor. The scattered water drop style drop lights in the stair case seem to break the tranquil water surface under the staircase. The little girl in the painting is holding a lotus leaf. It seems like she is afraid her clothes might get wet by the water.

The second floor office space applies the warm color tone of the ashtree wood to convey the kindness and unity of people working here. The modern office space makes use of China black marble to neutralize the frivolous feel of the white color tone. The whole space conveys to people modern and concise feel, with some primitive atmosphere.

049

开山设计顾问办公室
Office for KSEN Design Consultants

设计单位：厦门开山设计顾问有限公司
设计师：郭坤仲
项目地点：福建省厦门市
项目面积：700 m²
主要材料：石膏板、水泥自流平、实木线条、烤漆铁板
摄影师：刘腾飞

Design Company: Xiamen KSEN Design Consultants, Co., Ltd.
Designer: Guo Kunzhong
Project Location: Xiamen in Fujian Province
Project Area: 700 m²
Major Materials: plasterboard, cement, wooden lining, lacquered iron plate
Photographer: Liu Tengfei

由于本案的原始形态为老厂房，房顶有着足够的高度，这就为设计提供了更多的可能性。在具体的设计过程中，保留了粗糙的墙面，还加入了两个盒子状的空间造型，盒子的开口不仅能使各个空间得到巧妙的分割，还保证了各区域的隐秘性。同时又能很好地把室外的光线引入工作区域，使之形成一个隔而不断、分而不离的办公空间，使工作环境更加惬意，空间层次也丰富了起来，功能更加的多元化。

中国汉字的创意造型、木石铁泥沙的运用、极富传统情调的茶具和明式圈椅等设计元素在新与旧、粗与细、高与矮的对比中显得和谐而趣味盎然。白色是空间的主基调，配合黑色线条的收边及墙面灯带的装饰，勾勒出明快、幽雅的空间气氛，这样一个"静谧"而"优雅"的场所，可以使人身心放松，这也是现代办公环境所应具有的新特性。

The original form of this project is an old factory, with sufficient height of the ceiling which provides more possibilities for the design. During the concrete design process, the design retains the coarse wall surface, accompanied with two box style spaces, which openings not only acquire skillful division for the various spaces, but also guarantee the privacy of each space. The exterior lights are well introduced inside the work space, producing an office space as a whole while possessing separations, which is more cozy and has more layers, with diversification in functions.

Design elements such as creative format of Chinese calligraphy, application of wood stone iron sand, tea wares of traditional artistic conceptions and round-backed armchairs produce harmony and great interests in the contrast of old and new, thin and thick, high and low. As the tone color for the space, white color is accompanied with black lining and the decoration of wall light belts, picturing some brisk and elegant space atmosphere. This tranquil and graceful location would make people feel relaxed, as a new feature for modern office environment.

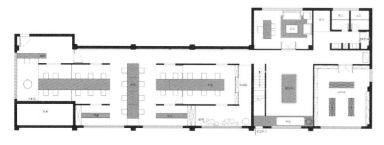

易·空间办公设计

Yi · Space Office

设计单位：福州易道装饰设计有限公司
设 计 师：曾昊
项目地点：福建省福州市
项目面积：248 m²
摄 影 师：周跃东

Design Company: Fuzhou Yidao Decorative Design Co., Ltd.
Designer: Zeng Hao
Project Location: Fuzhou in Fujian Province
Project Area: 248 m²
Photographer: Zhou Yuedong

这是一套办公空间的设计，集公共办公、会议、商务洽谈、休闲于一体。在设计过程中考虑的是如何将现代办公的时尚元素融入于源远流长的传统文化中，让传统的中式空间变得时尚起来，以激发起愉悦的情绪，同时又使得现代的空间多一份传统的韵味与高贵。因此，"易"就成为了空间的主题。

候客大厅的设计大胆地采用了镜面与不锈钢的材质，并把拆分过的汉字偏旁部首与水墨画通过局部的灯光予以渲染，吊顶的块面化处理强化了空间的纵横感，使人们在这个空间中体验到了一种新奇的感受。

会客区的设计意图是扭转过于理性与严肃的会客氛围，着重打造休闲与放松的空间韵味。在这里，沙发背景墙的纯朴肌理与不锈钢构件形成了视觉上的反差，又彼此和谐地交融在一起。宽松的沙发造型、典雅的花鸟图案壁纸，都使空间变得活泼起来。

休闲的设计思想同样体现在员工的办公区域内及阳台的设计中，有着流畅曲线和缤纷色彩的椅子与灯具成为这里的主角，舒缓了办公空间带给人们的紧张感。

Custom Office Space

This project involves an office space, integrating public office, conference, business negotiation and leisure as a whole. During the design process, the designer needs to consider how to integrate the fashionable elements of modern office inside the long-standing and well-established traditional culture, making the traditional Chinese space become fashionable, arousing pleasing mentalities and making the modern space possess some traditional charms and nobility, thus arising the theme of the space, "Yi".

The design of the waiting hall boldly applies mirror surface and stainless steel materials. The split radicals of Chinese characters and ink paintings apply colors through lights to some parts. The block and surface treatment of the ceiling strengthens the vertical and horizontal feel of the space, making people feel some novel sense inside the space.

The design intention of the drawing room is to change the much too rational and serious atmosphere, emphasizing on the creation of leisure and relacing space charms. Here, the primitive texture of the sofa background wall and the stainless steel structures form some visual contrast, while integrated with each other harmoniously. The cozy sofa format and the elegant wallpaper of flower and bird patterns all live up the whole space.

The leisure design concept is also represented in the staff office area and the balcony. The chairs and lighting accessories with smooth curves and colorful tone become the protagonist here, softening the sense of tension brought by the office space.

太空利器 Space Edge Tool

设计单位：谷腾设计师事务所
设 计 师：谷腾
项目地点：上海市

Design Company: Guteng567 Design Firm
Designer: Gu Teng
Project Location: Shanghai

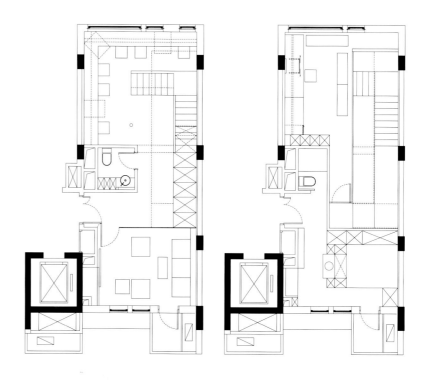

Custom Office Space

此空间的格局较之前的原始布局进行了较大的改造。由于空间面积比较小,所以特意在设计上利用一些材质的特性实现了很多功能的需求,在满足使用功能的基础上,将视觉上的观感尽量放大,运用了镜面和玻璃的特性使空间有延伸感,并且也有了趣味性。

二层墙体与玻璃构成的斜线相结合,使空间的体量加大,而且对于没有围栏的过道也起到了保护的作用。家具的设计采用了一种元素,且形态多变,使之在视觉上有更加统一的观感。

The space layout proceeds with great renovations from the original ones. As the space is kind of small, the designer intends to make use of the nature of some materials in the design to realize many functional requirements, while meeting with the functions, maximizing the visual scale. The application of mirror surface and glass extends the space, displaying more interests.

The second floor is combined with the glass slanting lines, extending the space scale, performing some protective function for the corridor with no fences. The furniture design applies one element with variations, producing some consistent vision in the visual sphere.

中企绿色总部·广佛基地办公室
ZhongQi green headquarters · guangfo base office

设计单位：广州共生形态工程设计有限公司
　　　　　WWW.COCOPRO.CN
项目地点：广东省佛山市
项目面积：800 m²
主要材料：大理石、复合实木地板、
黑色镜面不锈钢

Design Company: Guangzhou C&C Design Co., Ltd. / WWW.COCOPRO.CN
Project Location: Foshan in Guangdong Province
Project Area: 800 m²
Major Materials: marble, composite solid wood floor, black mirror surface stainless steel

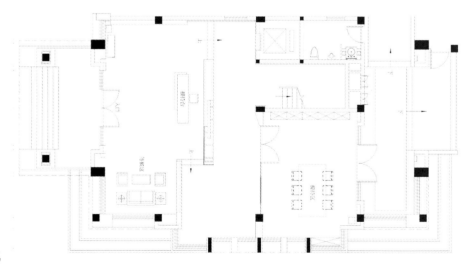

Custom Office Space

中企绿色总部·广佛基地位于广佛核心区域——佛山市南海区里水镇东部，总占地面积30万平方米，建筑面积50多万平方米。该项目由生态型独栋写字楼、LOFT办公、公寓、五星级酒店、商务会所、休闲商业街等组成。

本案突出"office park"和"business casual"的现代办公理念，办公环境拥有良好的通风和光照，以及花园式的生态环境，以实现工作环境中人与自然的和谐统一。在工作中，女士可以不穿高跟鞋、套装，而选择具有休闲气质的小西装、平底鞋。男士可以不系领带，选择白衬衣、休闲西裤、休闲鞋，但是不能穿运动鞋。希望倡导一种"面对面"的工作理念，提倡"沟通、交流和互动"的工作方式，并希望办公人员能有一种亲切的归属感，在轻松、愉悦的气氛下互动。

本案亦是在这种办公理念的倡导下，以人为本，将传统办公空间的沉稳、严肃感与生活空间的休闲、愉悦气息结合起来，创造出一种办公空间独有的休闲格调。

This project is located in Guangfo's core area, to the east of Lishui Town, Foshan's Nanhai District, with total plot area of 300000 m² and total building area of more than 500000 m². This project is composed of ecological single office building, LOFT office, apartment, five-star hotel, business club and leisure business street.

This project highlights the modern office concept of "office park" and "business casual." The office environment boasts fine ventilation and lighting, together with garden-style ecological environment, attaining

harmony of man and nature in the work environment. During work, the ladies do not have to wear high-heeled shoes or suits, but small suits and flats of leisurely temperament. The gentlemen do not have to wear ties, but white shirts, casual trousers and casual shoes, but no sneakers. This project hopes to advocate some "face to face" work concept, promoting some work style of "negotiation, communication and interaction," thus the staff can enjoy some intimate sense of belonging and proceed with interaction under relaxing and pleasing atmosphere.

This project is also guided under this office concept, with human-oriented principle, combining the sedate and solemn feel of traditional office space with the leisurely and pleasing atmosphere of life space, creating some leisurely tone exclusively to office space.

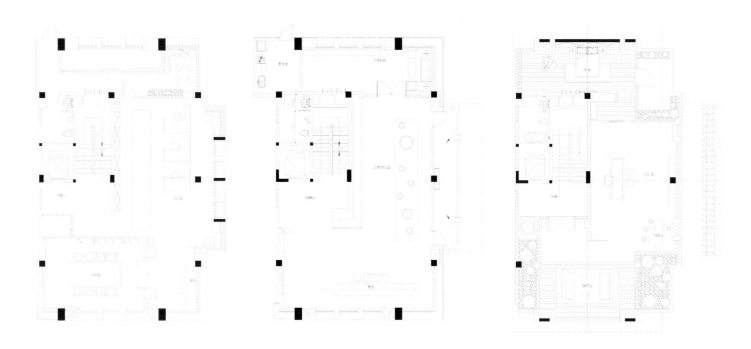

卓越世纪中心办公式样板房
Excellence Century Center Office Style Sample House

设计单位：深圳创域设计有限公司
设 计 师：殷艳明
项目地点：广东省深圳市
项目面积：120 m²
开 发 商：深圳世纪城房地产开发有限公司

Design Company: Shenzhen Creative Space Design Co., Ltd.
Designer: Yin Yanming
Project Location: Shenzhen in Guangdong Province
Project Area: 120 m²
Developer: Shenzhen Century Town Real Estate Development Co., Ltd.

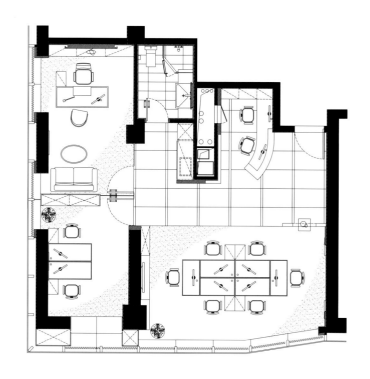

Custom Office Space 075

本办公项目的面积为120 m²，为咨询公司的办公场所，在充分考虑服务对象特点的情况下，设计师细致地梳理了潜在客户的各种使用需求，合理地对空间进行了功能分区，使空间的利用率最大化，其设计体现出了中国人常讲的一句话"麻雀虽小，五脏俱全"。

本设计方案的色彩以米色、咖啡色为主，在色彩块面的边缘饰以黑色的线条进行勾勒，一如绘画时采用的线描手法。材料的选择则以具有自然肌理效果的材质为主，与整体色彩的搭配相得益彰，使整个空间透露出成熟、稳重和轻松舒适的气息。

设计师在会议室的构思上独具匠心：首先以开敞的方式形成了一个多功能的空间，会议桌旁的一面白墙完全成为其展现思想的平台，屋顶顶棚以连续的斜面序列造型配以柔和的间接光源，营造出开放、友好、和谐的空间表情。

The area for this office space is 120 m², as an office location for a consulting corporation. While fully considering the situation of the service object, the designer deliberately analyzed the various potential practical requirements of the customer, carrying out functional division of the space in an appropriate way, and maximizing the use ratio of the space, with design representing a saying of Chinese: Although the sparrow is quite little, the internal organs are quite complete.

The designer is quite ingenious in the conception for the conference room. First, he proposed a multi-functional room with open style, with the side white wall beside the conference table becoming a platform demonstrating the thinkings. With consecutive beveling orders, the roof ceiling format is accompanied with soft indirect lights, creating some open, friendly and harmonious space expressions.

The design project focuses on beige and coffee colors, decorated with black lining on the edge of the color blocks and surfaces, just like the line drawing approach in drawings. The materials selection focuses on materials of natural texture effects, with the whole colors bringing out the best in each other, making the whole space send out some mature, sedate and relaxing atmosphere.

HED+OFFICE

设计单位：LA.H 贺钱威室内设计有限公司
项目地点：浙江省宁波市
项目面积：300 m²
主要材料：素水泥、锈板砖、大理石、钢材、镜面

Design Company: LA.H HED Interior Design Co., Ltd.
Project Location: Ningbo in Zhejiang Province
Project Area: 300 m²
Major Materials: plain cement, rust plank brick, marble, steel, mirror surface steel

HED 是一家专业设计公司，室内是 6 m 的挑高空间。一层为门厅、接待区、创意一区、经理和财务办公室、吧台、卫生间。

三层墙面为 5.5 m 的大落地玻璃墙，可以看到窗外的绿化带和小河，户外景观得天独厚。设计师遵循室内、外相互呼应的设计原则，充分利用户外景观。在室内空间设计上充分利用自然光，材质选择上呼应户外，利用天然的环保绿色材料，大面积应用素水泥墙与大面积留白，来映衬户外的自然景观，让绿色与阳光充分渗透到室内的每一个角落，仿佛户外与室内的界面已不存在，让每个人都最大限度地感受绿色、感受大自然。

在空间处理上，设计师将禅意空间的静态之美营造出来，光影的美、建筑的美、质朴的美、和谐的美，真正实现了"以人为本"的设计理念。

二层在空间上处理成错落有序的错层空间，以抬高的 L 形走道为分割界面，区隔出创意二区、会议洽谈区、材料样板区、设计总监办公室。

HED is a professional design company, with 6 m high space in the interior. The first floor includes hallway, reception area, innovation area 1, executive and financial office, bar counter and wash room.

The second floor forms a well-proportioned split-level space, with uplifted L shape corridor as the separating surface, producing innovative area 2, conference and negotiation area, materials show space and design director's office.

The third floor attains 5.5 m grand French window style glass wall, through which one can see the greening belt and river outside, possessing exclusive exterior landscape. The designer follows the design principle of interior and exterior corresponding with each other, while making full use of the outdoor landscape. The interior design makes full use of natural light, with materials echoing the outdoor space. The natural environmental green materials apply a large area of plain cement wall and blank space design reflecting the exterior natural views. Green and lights can get into every corner of the interior space, seeming like the interface of outdoor and interior space does not exist any longer. Thus every one can enjoy as much green and nature as possible.

In space treatment, the designer displays the tranquil beauty of the zen space. The beauty of light and shadow, the beauty of building, primitive beauty and harmonious beauty all truly accomplish the human oriented design concept.

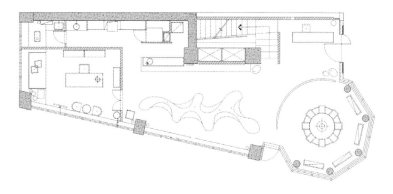
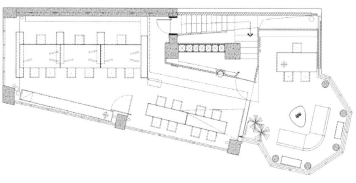

万科钻石广场 Loft A4 办公样板房
Vanke Diamond Square Loft A4 Office Model House

设计单位：深圳创域设计有限公司
设 计 师：殷艳明
项目地点：四川省成都市
项目面积：120 m²

Design Company: Shenzhen Creative Space Design and Decoration Co., Ltd.
Designer: Yin Yanming
Project Location: Chengdu in Sichuan Province
Project Area: 120 m²

自然万物有着不一样的形态、生命与节奏，本案的设计理念来源于对自然的向往与追求。自然是一种无形、抽象的事物，所以设计师在本案的设计中采用了不规则的形状与块面变化来诠释大自然的美。

原始的室内空间相对方正，设计师通过曲折、斜向及切割等方式来体现空间关系的变化，使空间产生一种具有韵律的节奏感。从入口至前厅，一个斜向的切割打破了原有空间的沉重感，使得空间更具活力与动感。前厅与办公区采用大切割的手法打破了整齐划一的空间色调，体现出结构的残缺之美。中空的吊高作为垂直空间的移动动线，形成了一个新颖、独特的 Loft 空间。

二楼的高管办公区与会议室（创意空间）不分区，便于整个团队成员的相互沟通与协调，有助于增强团队的凝聚力。在色彩上以黑、白、灰为主调，使空间在颜色上有深浅对比，强调时尚感、空间感及设计感。局部结合一些跳跃的绿色调，为空间注入一份休闲的气息。绿色的点缀活跃了整个空间的氛围，为工作与生活之间找到了平衡点，呼应了设计的主题。

Custom Office Space 085

087

Everything in nature has different format, life and rhythm. The design concept of this project originates from longing and pursuit of the grand nature. Nature is some intangible and abstract matter, thus the designer applies some form and block with irregular format in this project to interpret the beauty of nature.

The original interior space is comparatively square and the designer displays the variation of space relationship through approaches of curves, slanting and cutting, etc., to create some rhythmic feel inside the space. From the entrance to the vestibule, a slanting cutting breaks through the heavy feel of the original space, entrusting the space with more energy and vigor. The vestibule and the office applies grand cutting approach to break through the integral and consistent space color tone, representing the fragmentary beauty of the structure. As a moving line in this vertical space, the cavity hanging produces a novel and peculiar loft space.

The senior executive office space and the conference room "innovation space" on the second floor are not separated from each other, convenient for the mutual negotiation and communication among the team members, contributing to strengthen the cohesion of the team. The tone colors focus on black, white and gray, producing dark and light contrast among the space colors, emphasizing on fashion feel, space feel and design feel. The bouncing green colors on some parts instill some leisurely atmosphere into the space. The ornament of green colors activates the atmosphere of the whole space, finding out a balance point between work and life and echoing the design theme.

新创广告办公室

Office for Xinchuang Advertising

设计单位：福州中和设计事务所
设 计 师：陈锐锋
项目地点：福建省福州市
项目面积：189 m²
主要材料：白色地砖、灰木纹大理石、灰镜

Design Company: Fuzhou Zhonghe Design Firm
Designer: Chen Ruifeng
Project Location: Fuzhou in Fujian Province
Project Area: 189 m²
Major Materials: white floor tile, gray wood grain marble, gray mirror

办公室装修的布局、通风、采光、人流线路、色调等的设计恰当与否，对工作人员的精神状态及工作效率影响很大，而办公室是脑力劳动的场所。因此，办公室装修应该更加重视办公室内的环境营造，借以活跃人们的思维，提高办公效率。同时，办公室也是企业整体形象的体现，一个完整、统一而美观的办公室形象，能增强客户的信任感，同时也能给员工以心理上的满足感。这些都是在做办公室装修时所应该考虑的。

本案作为广告创意公司，在设计上的定位是要具有东方气质，于是秦俑、宋体汉字、十二生肖……这些传统中国文化的代表符号就融合进了新创广告公司的设计主题中，汇集成了和谐而完整的东方文化主题空间，以简练的设计手法表达出创意型主题的办公空间。

Custom Office Space 091

The design of the office's layout, ventilation, lighting, streamline, color tone, etc. would have a great effect on people's spiritual status and work efficiency, and office is a place for brainwork. Thus, the office decoration would focus more on the environmental creation of the office interior space, thus activating people's thinking and uplifting office efficiency. While at the same time, the office is also a presentation of the corporate whole image. A complete, integral and nice office would strengthen the customers' sense of trust and provide the staff with psychological satisfaction. All these should be taken into considerations while decorating an office.

As an advertising creation company, the design should possess oriental temperament, thus some traditional Chinese cultural representative symbols such as Terra-cotta Figures of the Qin Dynasty, Song Typeface characters, 12 Chinese Zodiac signs are integrated into the design theme of this advertising company, performing harmonious and complete oriental cultural theme space, presenting the innovative theme office space with concise design approaches.

东方国际·创冠集团香港总部
Oriental International·C&G Group Hong Kong Headquarters

设计单位：厦门宽品设计顾问有限公司
设计师：李泷
项目地点：香港
项目面积：1500 m²
主要材料：灰色橡木、银貂大理石、镜面不锈钢、壁纸

Design Company: Xiamen Deep Design Consultant Co., Ltd.
Designer: Li Long
Project Location: Hong Kong
Project Area: 1500 m²
Major Materials: gray oak, silver marble, mirror surface stainless steel, wallpaper

本案的设计重点在于强化企业面向世界的国际化形象，并融合企业文化中以人为本的概念，在空间气质中体现高度的国际视野及人文关怀。

项目所在地的地域背景为本案提供了丰富的设计素材，它所具有的中西结合的殖民地文化特色催生出更多的设计可能性，人与人之间不同的相处方式及公司管理特性也使空间的规划有了不同于传统空间的气质。

公共空间的简约大气来源于精致材质的大面积铺陈及色调对比中所隐含的东方美学。该设计顺应企业在办公模式、管理方式及内部组织结构的变化，对空间设计提出了新的要求。设计师要理解办公空间是承载工作人员梦想的舞台，设计中要注重舒适度的营造。因此要与工作人员做好充分的沟通，并理性地进行思考。

Custom Office Space

The design of this project focuses on the internationalized image of the company, integrated with the human oriented concept in the corporate culture, while representing grand international vision and human care in the space temperament.

The conciseness and magnificence of the public space originates in the large area spreading of delicate materials and the oriental aesthetics implicit in the color contrast. The design follows the corporate variations in office modes, management modes and interior organizational structures, putting forward new requirements for the space design. The designer should understand that office space is a stage carrying the dreams of the staff and the design should focus on the creation of comfort. Thus, the designer should have sufficient communication with the staff and carry out thinking in a rational way.

The local background of the project provides the design with abundant design materials, with its colonial cultural features integrating east and west producing more design possibilities. The getting along method among people and the corporate management features allow the space planning to have some temperament different from traditional space.

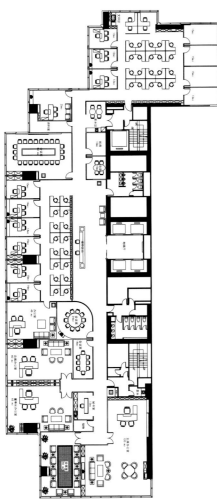

办公空间的光影魔术
Light and Shadow Magic of Office Space

设计单位：东道设计有限公司
设 计 师：李川道
项目地点：福建省福州市
项目面积：260 m²
摄 影 师：申强

Design Company: Dongdao Design Co., Ltd.
Designer: Li Chuandao
Project Location: Fuzhou in Fujian Province
Project Area: 260 m²
Photographer: Shen Qiang

本案是东道设计有限公司对其办公空间的改造案例，设计师坚持"以人为本"的核心价值，利用不同的材质、搭配适宜的光照，营造出了现代、时尚、充满动感的办公环境。对设计者而言，作品最成功的地方莫过于在可以满足其办公需求的情况下超越功能本身，带来更多有形和无形的附加值。

休闲室的设置是人性化办公的体现，在工作之余可以再次放松心情、增强同事之间的沟通与了解，劳累的心情即可烟消云散。这种对于人性的尊重可以使员工有一种归属感，对工作的热情也会随之被激发出来。

设计突出强调材料本身的质感，无论是平整的玻璃、粗糙的隔墙还是具有手工质感的木块，透过其自身的比例关系将原本产生冲突的空间与材料转化成空间的主角，低调地表达着空间的生命核心。

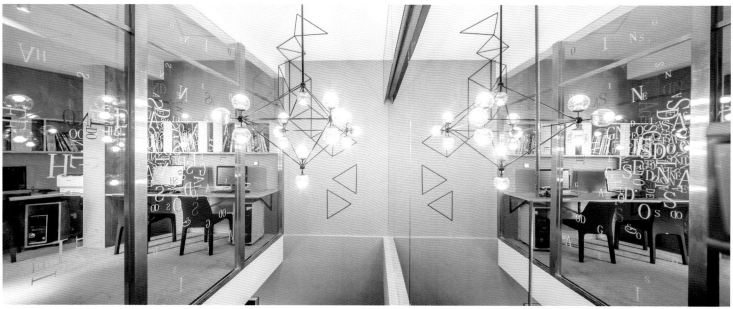

This project is the regeneration project of Dongdao Design Co. towards its office space. The designer insists on the core values of human oriented principle and produces a modern, fashionable and dynamic office environment with different materials accompanied with proper lighting. The most successful part of the work lies in that it can go beyond the function itself and meet with the office requirements, creating more tangible and intangible additional values.

The design highlights the texture of the material itself. For smooth glass, coarse partition wall and wood block of handmade texture, the self proportional relationship transforms the original space and materials with conflicts into the protagonist of the space, expressing the space core in a low-profile way.

The leisure space setting is the presentation of human office. When you are free from work, you can relax the moods, strengthen the communication and understanding among colleagues and the exhausted feel would disappear. This human respect would entrust the staff with some sense of belonging, while enthusiasm for work would be brought out.

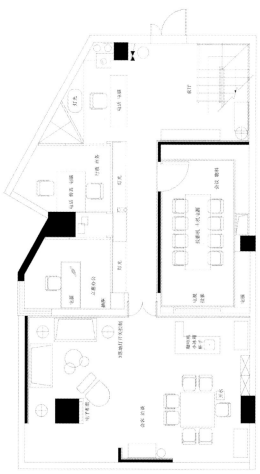

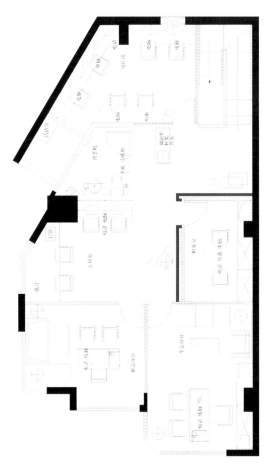

鸿隆世纪广场 A+B 户型办公样板间
Hong Long Century Plaza, A+B Office Model Room

设计单位：深圳创域设计有限公司
设 计 师：尹艳明
项目地点：广东省深圳市
项目面积：503 m²

Design Company: Shenzhen Creative Space Design Co., Ltd.
Designer: Yin Yanming
Project Location: Shenzhen in Guangdong Province
Project Area: 503 m²

投资公司向来被人们认为是服务于高端客户的专业机构，所以在设计上就要求塑造出自信和值得信赖的公司形象。设计师把艺术气质和高雅的品位作为本设计的依据，表达出简练、考究、完美的空间气质。

虽然有503 m²的空间可供发挥，设计师却抛弃了传统的空间切割方式，甚至连经理室也没有封闭起来，而是保持了整个空间的开敞与流动，表达了投资公司开放、自信的从业态度。

在充分考虑空间功能需求的情况下，设计师通过对顶棚、墙面、地面的造型方式和不同材料的使用，将各个功能区自然地分割开来，形成相对独立的使用空间。

经理办公区域是设计师关注的重点，顶棚以连续相交的圆形打破了方形空间的束缚，成为视觉的焦点，办公桌后面的背景墙仿佛是一幅自然的画卷，稳重中有着较强的亲和力。会议室无论是天花、灯饰、会议桌、装饰画还是墙面的处理，均采用了方形，色彩以黑、白、咖啡和米黄相搭配，营造出相对严肃、理性的氛围。

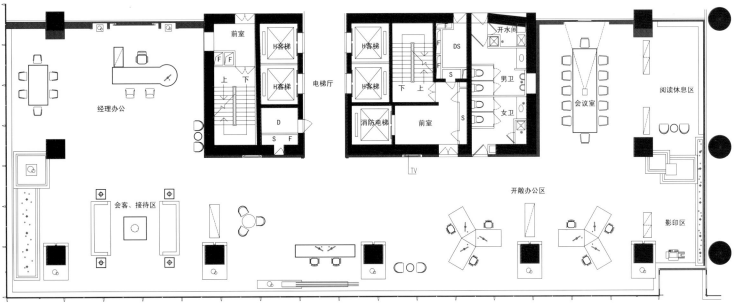

总体而言，整个空间设计强调空间的结构美与序列感，局部饰品的运用既体现了现代、时尚的审美情趣，也散发出自然的意境，起到了"画龙点睛"的作用。设计师还特别利用玻璃窗将室外的城市景观引入室内，让室外空间成为室内空间的背景，体现出城市绿洲的设计理念。

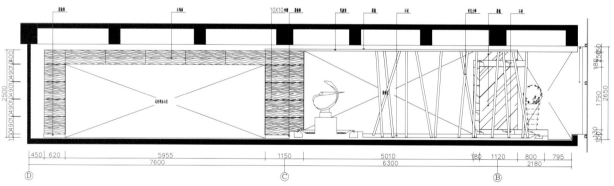

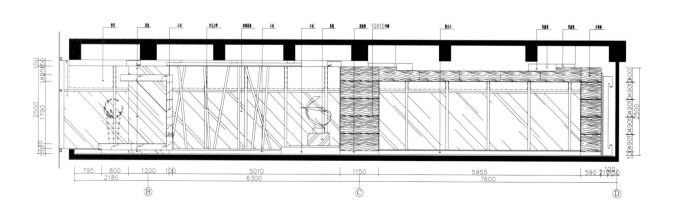

Investment companies have always been considered to be high-end customer service professional organization, so interior design should establish self-confident and trustworthy corporate image. We are deeply rooted in artistic temperament and exquisite taste, expressing concise, meticulous and perfect space atmosphere.

Although there is a 503 square meters-space available for play, the designers have abandoned the common space cutting manner, and even the manager's office is not closed up, while maintaining the open and flowing manner of the space, displaying the open and confident industrial attitudes of the corporation.

While taking into full considerations the space functional requirements, the designer naturally separates various functional areas through format approaches of ceiling, wall surface and floor and application of different materials.

The manager's office area is the focus of the designer. With consecutive round shapes, the ceiling breaks through the constraints of square spaces, becoming the visual focus. The background wall behind the office table is like a natural painting, with powerful intimate feeling in the sedateness. Inside the conference room, the ceiling, lighting accessories, conference table, decorative table and wall all apply square shapes, with color collocations of black, white, coffee and beige, producing solemn and rational atmosphere.

To sum up, the whole space design focuses on the structural beauty and order feel of the space, with application of ornaments on some parts not only representing modern and fashionable aesthetic interests, but also emitting some natural artistic conceptions, making the finishing point. The designer specially introduces exterior urban landscape inside with glass windows, making the exterior space become the background for the interior space, displaying the design concept of urban oasis.

Angular Momentum

Angular Momentum

设计单位：上海牧桓建筑 + 灯光设计顾问
设 计 师：赵牧桓
参与设计：赵玉玲、胡昕岳
项目地点：上海
项目面积：700 m^2
主要材料：密度板、冷烤漆、不锈钢、环氧树脂
摄 影 师：周宇贤

Design Company: MoHen Design International
Designer: Hank M. Chao
Associate Designers: Zhao Yuling, Hu Xinyue
Project Location: Shanghai
Project Area: 700 m^2
Major Materials: density board, cold stoving varnish, stainless steel, epoxy resin
Photographer: Zhou Yuxian

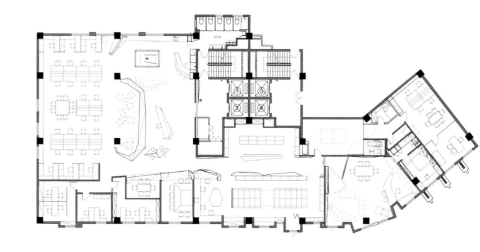

Custom Office Space

本案中，设计师试图通过向量来寻找空间流动的动能。

在平面布局上，设计师利用休息区的隔墙将入口的动线进行分流，将私密的大工作区与公共的开放区划分开，两侧分布着中小型会议室和台球室、娱乐室。

在设计理念上，设计师将不等分的二维平面构筑成三维的量体，形成一种凝结的空间动能。不同向量的平面切块互相交汇、连结，并在轴线上扭动翻转，宛如一个运动体在时间冻结时的定格，也就是说隔墙不再是立面上的一个维度，而是处于一种运动状态的静止画面。因此在进入空间后的视觉和空间感就有了改变，空间因此有了流动感，是一个视觉暗示，但却改变了人的实质感知力，空间也就有了不同的趣味特点。

会议室的分隔墙面用的是雕刻纹样的玻璃，并在里面安放灯带，以强调办公室的工作属性。图书资料室采用开放式的设计，里面并无明显的隔间，中间由一个长廊连接，可作为一个小型的短会交流区，走道直通主管办公室，并分流至娱乐健身房。

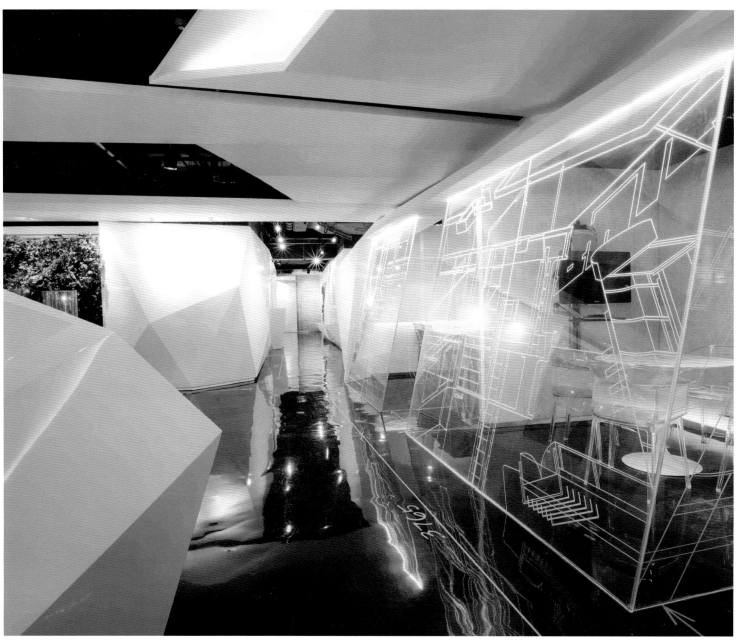

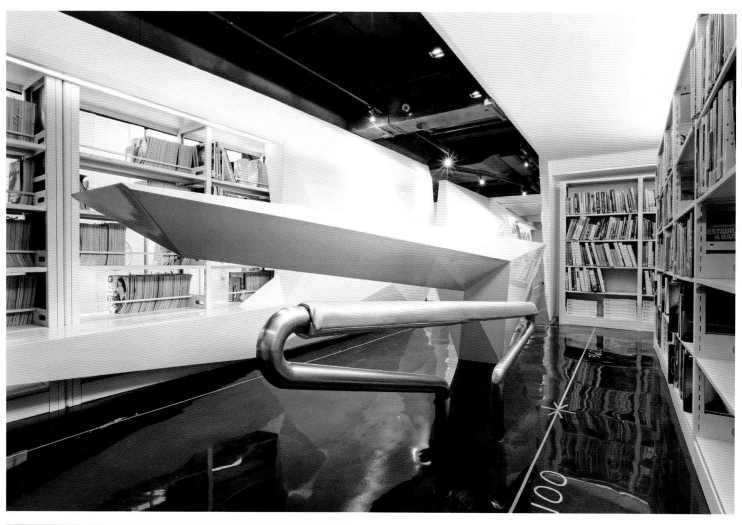

In this project, the designer tries to seek for the space flowing energy through vector quantity.

In plane layout, the designer makes use of the partition wall of the lounge area to proceed with stream separation of the entrance moving lines, separating the private grand work area from the public open area, with medium to small conference rooms, billiard parlor and entertainment space on both sides.

The separation wall of the conference room applies glass of carving pattern, with light belt inside, thus stressing the work nature of the office. The library and reference room applies open-style design, with no obvious separation inside and the middle part is connected with a long corridor, which can be used as a tiny short meeting communication area. The corridor leads directly to the supervisor's office room, with separate lines leading to the entertainment and gymnasium.

In design concept, the designer constructs three-dimensional body mass with unequal plane, forming a condensed space energy. The space blocks of different vector quantities are interlaced and interconnected with each other, twisted and overturned in the axis, which is like the freeze-frame of a moving body when time stops. It can also be said that the separation wall is no longer a dimensionality on the facade, but a static picture in moving situations. Thus, the vision and space feels after entering the space are various, creating flowing feel for the space. As a visual implication, this changes people's substantial perceptivity, with the space acquiring some different charms.

前线共和广告公司办公室
Office for Frontline Republican Advertising Office

设计单位：福建国广一叶建筑装饰设计工程有限公司

方案审定：叶斌

设 计 师：何华武、龚志强、蔡秋娇、杨尚炜

项目地点：福建省福州市

项目面积：1000 m²

Design Company: Fujian Guoguangyiye Architectural Decorative Design and Engineering Co., Ltd.
Project Examiner: Ye Bin
Designers: He Huawu, Gong Zhiqiang, Cai Qiujiao, Yang Shangwei
Project Location: Fuzhou in Fujian Province
Project Area: 1000 m²

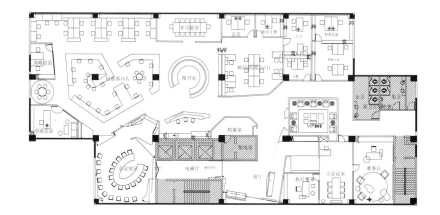

优质的空间设计其实并无确定的标准,因为对现在的人而言,优质已不仅只是表象的展示,而是一种态度。主题为"Ice change"的主题办公室意为"冰变"的意思,是前线地产机构的办公室。大胆的造型、鲜活的陈设,围绕着明快、硬朗的体面造型,使其不仅仅局限于一般意义上的理论构筑,而是在变通之间超越了传统与现代的概念性对峙,使得空间成为有节奏感的办公环境。

进入前线地产机构的办公空间,一幅精致而梦幻的画面展现在我们眼前。前台的白色台面呈现出不规则的切割形态,背景墙则以暗色调铺陈,独特的肌理与色彩的反差吸引了人们的视线。与此同时,灯光从走道至前台开始加强,并在台面上增加了红色的光晕,鲜明地提示了这个功能区域的存在。前台这种独具匠心的设计让人产生一种渐进的视觉效果,并传递着惊喜。

休闲区是一个白色围合形的空间,其上方的灯饰也以圆形作呼应,是一处设计的亮点。这个区域的内部可供人们休闲与阅读之用,外围的方格可以放置书籍。与之配套的黑色吧台采用了几何造型,体量感十足。

开放式办公空间的布局使得视线可以自由地游走,设计师利用空间的布局来模糊工作与休闲的临界点,让IN与OFF的概念得到巧妙的切换。工作区的红色几何灯盒与白色灯光的塑造,与几何办公桌形成了呼应,清晰地规划出这个功能区域。

In fact, there is no exact standard for space of high quality, because for people today, quality is no longer superficial display, but an attitude. The theme of "Ice change" means change like ice, in the office of Frontline real estate agencies, bold design and vivid furnishings around the surface sense with bright and tough theme making the space beyond the conceptual confrontation between tradition and modern during flexibility, rather than bundle to the general sense of the theoretical architecture, making the space become the office environment with a sense of rhythm.

An exquisite and fantasy picture is displayed in front of us when we walk into the office of Frontline real estate agencies. The white table of Front Desk shows the irregular cutting shape, while the background wall is created with dark tone. The contrast of unique texture and color attracts people's attention. Meanwhile, The light strengthens from the walkway to the front desk, and red halo is added above the table indicating clearly the existence of this functional area. As a prelude to an office space, the front desk with ingenious design makes people produce the gradual visual effect, and passes on surprise.

As a highlight of the company building, Recreational Areas has a white co-shaped space, and the lights above are in the shape of round to echo. People relax and read within this region, and the books and

materials can be put in the external grid. The corresponding black bar with a geometric style is full of dimensional sense.

The layout of open office space allows the sight move freely. Using of space layout blurs the critical point of the work and leisure, and makes IN and OFF switch cleverly. The red geometric light box and the white light of the workspace form the echo with geometric desk, clearly planning this functional area.

树林里的办公室 | Office in the Forest

设计公司：深圳市昊泽空间设计有限公司
设 计 师：韩松
项目地点：广东省东莞市
项目面积：100 m²
主要材料：烤漆玻璃、仿真绿植、壁纸、石材、实木

Design Company: Shenzhen Haoze Space Design Co., Ltd.
Designer: Han Song
Project Location: Dongguan in Guangdong Province
Project Area: 100 m²
Major Materials: lacquered glass, simulated green plants, wallpaper, stone, solid wood

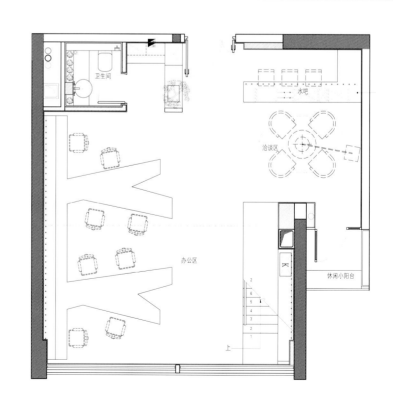

有一种感动,如春日的新绿悄然生长,融合于大自然中,这样让人惊喜的生命让梦想肆意生长,使疲惫的身心远离闹市,从而回归自我。

本案的设计定位为商用LOFT办公空间,虚拟客户为年轻的创业者。随性且富有趣味的空间布局形态赋予了空间具有灵动感的生命。

在功能划分上,一层为办公区、休息接待区,二层为个人办公区。一层的空间临窗,将窗外的绿色田园情趣引进来,使室内充满了生机与活力。一层保留了挑空的区域,实现了一、二层的相互沟通。二层的每一个独立办公室都有属于自己的一片绿色,它们相对独立但又与整个空间巧妙融合,绿色与白色的相互结合使空间的整体风格简洁、清新、自然且具生命力。

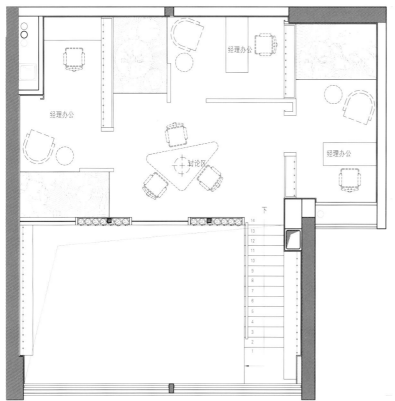

Some kind of moving affection is like the new green of the spring time growing quietly and integrated in the nature. This surprising life makes the dreams develop willfully and allows the tired body and heart to be far away from the noisy city, back to the original self.

The design of this project is positioned to be LOFT office space, with young entrepreneurs as the virtual customers. Casual space layout format entrusts the space with dynamic life.

In functional distribution, the first floor is office and lounge reception area and the second floor is personal office space. The window of the first floor office introduces the outside green pastoral interests inside, filling the space with vigor and vitality. The first floor retains the high area, attaining the mutual communication between first and second floors. Every independent office space on the second floor has its exclusive green space, mutually independent while ingeniously integrating with the whole space. The mutual combination of green and white makes the integral style of the whole space concise, fresh, natural and full of vigor.

唯知唯美

Purely Aesthetic Space

设计单位：福州创享空间设计
设 计 师：陈严
项目地点：福建省福州市
项目面积：130 m²
摄 影 师：周跃东

Design Company: Fuzhou Chuangxiang Space Design
Designer: Chen Yan
Project Location: Fuzhou in Fujian Province
Project Area: 130 m²
Photographer: Zhou Yuedong

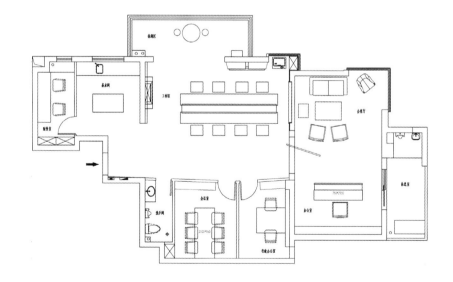

在本案的设计过程中,设计师遵循"将空间美感与功能需求相结合"的设计原则,在满足办公需求的同时,通过构建个性化空间,打造出充满魅力的办公环境。在本案中,设计师利用不规则元素,以简洁、大方、自然的设计手法充分表达了公司的企业文化特征。

在功能区域的分布上,设计师将工作区域分为开放与封闭两种模式,整个分区既能满足各部门的功能需求,同时又能对人员进行有效分流,并且使各部门之间能够有效沟通。

在公共区域的处理上,以简约、现代的设计手法,充分体现了实用与装饰相结合的设计理念。在空间色彩的应用上,以白色和木色为主色调,有效地提升了整个办公区域的品位。

在材料的选择上,设计师利用白色水泥漆作为主要的材料,搭配具有天然纹理的木质材料,再摆放上色彩鲜明、造型多样的工艺品做点缀,使整个空间传达出一份独特的自然氛围。

在隔断的设计上,设计师以白色块面的墙体作为隔断,不但形成了相对独立的办公区域,同时还营造出了干净纯粹的空间感。在色彩、材质及灯光的搭配下,整个空间更加明快、现代,使人产生一种愉悦感。

同时,设计师将强烈的体块感同灵动的平面流线元素有机地融合,运用白色块面墙体、天然木质材料及明快的流线形式,融合入户区的折叠造型,不仅增强了空间的视觉冲击力,而且在灯光的作用下,块状的立面产生出独特的光影效果,凸显出了空间的和谐感。

During the design process, the designer follows the design principle of the integration of space aesthetic feel and functional requirements. While meeting with the office requirements, through constructing a peculiar space, the design produces an office environment full of charms. For this project, the designer makes use of irregular elements and represents the corporate cultural features with concise, grand and natural design approaches.

In the layout of functional areas, the designer divides the working area into open and close two modes, with the whole area division which can not only meet with the functional requirements of different departments, but also achieve efficient distribution for the streams of people, accomplishing fine communication among them.

In the treatment of the public area, the designer fully represents the design concept of the combination of practicality and decorations through concise and modern design approaches. The space colors center on white and wood color as the tone, efficiently uplifting the taste of the whole office area.

In the selection of materials, the designer applies white cement paint as the major materials,accompanied with wood materials of natural texture and ornaments of artistic objects of bright colors and various formats, making the whole space display some peculiar natural atmosphere.

As for the design of partition, the designer makes use of white block wall as the partition, not only forming some comparatively independent office area, but also displaying some clear and pure space feel. With the collocation of colors, materials and lights, the whole space becomes more brisk and modern, arousing some pleasant feelings in people's heart.

While at the same time, the designer attains organic integration between intensive block feel and dynamic plane streamline elements, applying white block wall, natural wood materials and brisk streamline format, integrating into the folded format, which not only strengthens the visual impact of the space, but also makes the block facade display some spectacular light and shadow effects with the lights, highlighting the harmonious feel of the whole space.

道和设计机构办公室 | Office for Daohe Design Institution

设计单位：道和设计机构
设 计 师：高雄
参与设计：吴运棕
项目地点：福建省福州市
项目面积：90 m²
摄 影 师：施凯、李玲玉

Design Company: Daohe Design Institution
Designer: Kaohsiung, Taiwan of China
Associate Designer: Gao Xiong
Project Location: Fuzhou in Fujian Province
Project Area: 90 m²
Photographers: Shi Kai, Li Lingyu

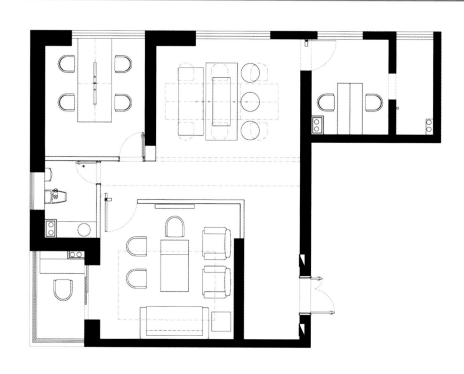

Custom Office Space

设计师的工作具有很强的创造性，对于自身的办公环境有着更为独特的需求。本案为道和设计机构的办公室设计，风格以现代简约为主。整个空间从功能组合、材料运用及色彩搭配上，都体现了现代实用、简洁明了的特点，既能满足办公需求，又具有装饰美感。

在材料选用上，最引人瞩目的是镜子与玻璃的运用，玻璃给人以通透感，传达出设计师工作中享受孤独的沉静感。镜子的使用使光影找到了跳跃的舞台，是空间体、面、光、影相互交织、相互沟通的最佳载体。

平面布局上，设计师根据办公室的功能需求将空间分为工作区、会议区及接待区，各个部分之间的联系紧凑而有序。装饰画、现代雕塑和圆凳等物品的陈设都凸显出了设计师不俗的审美品位及对办公环境较高的要求。

The designer's work has strong creative nature and has peculiar requirements for the office environment. This project is the office design for Daohe Design Institution, with style focused on modern conciseness. From functional grouping, materials application and color collocation, the whole space displays modern, practical, concise and clear features, meeting with office requirements, while with decorative aesthetic feel.

In materials selection, the most impressive part is the application of mirror and glass. The transparent glass conveys the sedate feel of the designer enjoying solitude during work. The mirror allowing the light and shadow to find the dancing stage, as the best carrier for the interlacing and inter-communication of the scale, surface and light, shadow of the space.

In plane layout, the designer divides the space into work zone, conference zone and reception zone according to the functional requirements of the office, with the connection of various parts terse and orderly. The layout of decorative painting, modern sculpture and stool highlights the uncommon aesthetic taste of the designer and the high requirements towards office environment.

世纪嘉美办公室 / Shiji Jiamei Office

设计单位：深圳市世纪嘉美装饰设计工程有限公司
设 计 师：龚德成
项目地点：广东省深圳市

Design Company: Shenzhen Shiji Jiamei Decorative Design and Engineering Co., Ltd.
Designer: Gong Decheng
Project Location: Shenzhen in Guangdong Province

本案原建筑层高有5m，如单做一层，层高太大，有点浪费空间；如做两层来使用，原高度又不够大。由于办公空间在一层，所以设计师考虑向地面以下挖掘空间。向地下延伸了0.6m后，设计师从前台后部做了往下走的三步台阶，造成一种错层感，使得立面空间的层次丰富了起来。

本案户型属于狭长形，这是一个缺点，为了避免这一点，设计师在过道的一侧做了半开放式的洽谈区，用夹丝玻璃做屏风隔断，营造出半通透的视觉效果。过道的白色墙面用灯光及墙面装饰做点缀，营造出一种自然、休闲的氛围。

洽谈区墙面用中纤板做成肌理效果，以洗墙地灯营造出古朴的意境。通过洽谈区简洁、素雅的混凝土楼梯通往二层，可以看到楼梯下面的观赏鱼池，放入了几条红色的锦鲤，在光影的映衬下，整个空间充满了灵性与趣味。

The original building floor is 5 m high. If we make it one single floor, the height would be too big which is a waste of the space. If we make it two floors, the original height is not big enough. As the office space is on the first floor, thus the designer considers excavating space underground. After the space is extended 0.6 m underground, the designer created three-step staircase from the back of the reception desk downward, creating some sense of staggered floors, enriching the layers of the elevation space.

A short-come for the project is the long and narrow format. In order to avoid that, the designer set a semi-open negotiation area on one side of the corridor, with wire glass as the screen partition, creating some semi-transparent visual effects. The white wall of corridor uses lighting and wall decorations as the ornaments, producing some natural and leisurely atmosphere.

The texture of the wall of the negotiation area is made of medium density fiberboard, producing some primitive artistic concept with ground lamps. One can get to the second floor through the concise and elegant cement staircase, observing the ornamental fish pond under the stairs with 8 red fancy carps inside. Reflected by the light and shadow, the whole space is full of spiritual power and interests.

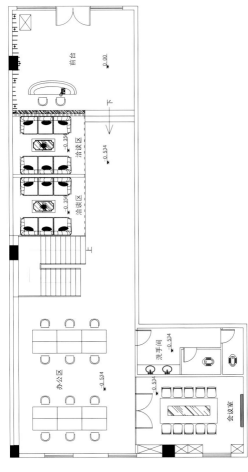

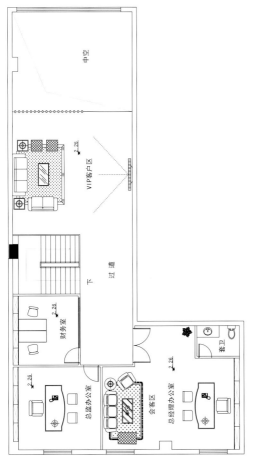

保发大厦劳伦斯珠宝写字楼
Baofa Building Lorenzo Jewelry Office Building

设计单位：深圳创域设计有限公司
设 计 师：殷艳明
项目地点：广东省深圳市
项目面积：1500 m²
主要材料：法国木纹石、黑镜钢、亚克力透光片、雅士白云石、软膜、指接板、皮革、金影木

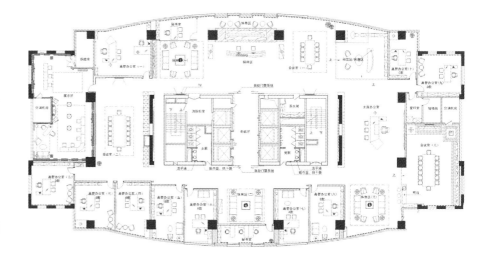

Design Company: Shenzhen Creative Space Design and Decoration Co., Ltd.
Designer: Yin Yanming
Project Location: Shenzhen in Guangdong Province
Project Area: 1500 m²
Major Materials: French wood grain stone, black mirror steel, acrylic glass, jazz white marble, mantle, wedge joint board, leather, wood

本案为劳伦斯 ENZO 中国区总裁及高层办公室，内部分为接待、会议、展厅及高管办公区，其在企业文化的展示中占有主要的角色。本案以新的观察角度与思考逻辑切入设计，以"钻石概念"为出发点，以钻石所寓意的"时尚、科技、创新、品质"作为空间设计定位，设计手法上突出主题，凸显定位。采用钻石切面的现代主义装饰手法，对"钻石"元素进行拆分、变化、提炼，采用明喻与暗喻的手法诠释空间的品质，使空间设计理念极具冲击力。

整个空间从电梯的入口出发向周围发散，形成流线完全开放的空间。空间主要分为两部分，一部分是开放的接待洽谈空间，另一部分为封闭的高层办公空间。这样分隔有利于空间的布局和交通的引导，在每个空间都设置休闲区作为缓冲，使办公室更加人性化。在设计手法上采用纵横切割的方式将原有空间进行分割并统一。地面与顶棚的设计元素上下呼应，在视觉上起到很好的延伸性，线面的结合使两者纵横交错，自然而然形成了一种横向与纵向关系上的视觉冲击力，增强了主体空间的节奏对比。

在色彩的选择上主要由背景的重色、主体的明亮色和点缀色构成。在造型设计中，设计师将整个劳伦斯企业文化融入于设计。时间廊的设计给通道赋予了新的活力，同时成为企业的文化展示墙。

This project is for ENZO China zone's presidential and high-level offices, with the interior part divided into reception area, conference area, exhibition hall and administrative office zone, which play a leading role in the presentation of the corporate culture. This project starts from new observation angle and thinking logics, with "diamond concept" as the starting point and its implied "fashion, technique, innovation and quality" as the space design orientation. The design approach highlights the themes and positioning. With modern decorative approach of diamond cutting surface, the design carries out separation, variation and subtracting towards the "diamond" elements and entrusts the space design concept with powerful impact force through interpreting the space quality with simile and metaphor.

The whole space radiates from the entrance of the elevator, forming wholly open streamline space. The space is mainly divided into two parts, one is the open reception and negotiation space and the other is the enclosed high-rise office space. This kind of separation is good for the space layout and the guidance of transportation, setting leisure

zone in every space as the buffer space, and making the space much more convenient. In design approach, the design applies vertical and horizontal cutting approach to carry out separation and integration of the original space. The design elements over the ground and the ceiling correspond with each other, extending the space visually while the combination of lines and surfaces making both interlaced with each other, naturally forming some visual impact in horizontal and vertical relationships, and strengthening the rhythmic contrast of the body space.

The color selection is mainly composed of dark colors in the background, the main part bright colors and decorative colors. In format design, the designer integrates the whole ENZO corporate culture inside the design. The design of the time corridor entrusts the passageway with some new energy, which becomes the corporate cultural presentation wall.

石油化工交易所办公空间
Office Space for Petrochemical Exchange

设计单位：深圳市萃成环境艺术设计有限公司
设 计 师：陈昆明
项目地点：广东省深圳市

Design Company: Shenzhen Cuicheng Environmental Art and Design Co., Ltd.
Designer: Chen Kunming
Project Location: Shenzhen in Guangdong Province

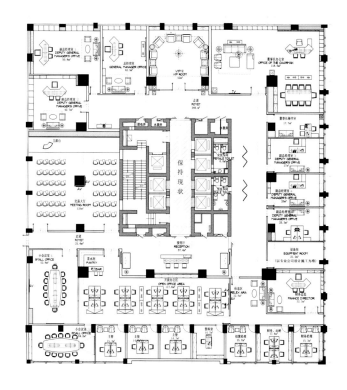

深圳石油化工交易所是深圳前海深港现代化服务业合作区中首批入驻的企业之一，交易所是一个立足中国、面向世界的集石化产品定价中心、交易中心、资讯中心、金融中心与供应链管理中心于一体的国家级、国际化的交易平台。

作为如此大型、新型的国际化办公空间，设计者不是一味地追求设计的繁琐及富丽堂皇的造型，而是通过和谐的色调、充足的光源、舒适而环保的材料，营造出简约而舒适明快的办公环境，让员工有"家"的感觉，从而提高工作效率，也使得企业有一个良性的发展。

本案整体空间色泽明快、设计手法简洁大方，相同的设计语言贯穿于整个空间中，为办公空间营造出一种安静、平和的环境。

在软装饰上采用了许多中国元素，比如董事长办公室中古典的红木雕花家具、中国瓷器等饰物，以及以中国画和书法为主的装饰挂画等，都体现出一家具有国际视野的现代企业的精神。设计理念和企业发展理念高度一致的融合，也使企业的文化氛围充分地体现出来，从而达到现代空间设计与中国文化的完美结合。

Shenzhen Petrochemical Exchange is one of the first batch of corporations entering Shenzhen-Hong Kong Cooperation Zone on Modern Service Industries in Qianhai Area. This exchange is a national and internationalized trading platform integrating petrochemicals pricing center, trading center, information center, financial center and supply chain management center based in China and facing the whole world.

For such grand and new international office space, the designer does not solely pursue the complexity of design and splendid format, but create concise, cozy and brisk office environment with harmonious color tone, sufficient light, comfortable and environmental materials, making the staff feel at home and promoting work efficiency, while maintaining sustainable development for the corporation.

The whole space has brisk colors and the design approach is concise and magnificent, with same design language running through the whole space, creating some quiet and peaceful environment for the office space.

The soft decorations apply many Chinese elements, such as ornaments of classical rosewood carving furniture and Chinese porcelain in the president's office and the decorative painting focuses on Chinese painting and calligraphy, all representing the modern corporate spirits of international views. The highly consistent integration of design concept and corporate developing concept fully displays the corporate cultural atmosphere, thus attaining the perfect combination of modern space design and Chinese culture.

波龙办公室 / Bolon Office

设计公司：玄武设计
设 计 师：黄书恒、许棕宣、陈昭月
项目地点：台湾省台北市
项目面积：245 m²
主要材料：波龙毯、壁纸、玻璃隔断
摄 影 师：王基守
撰　　文：程歆淳

Design Company: Sherwood Design Group
Designers: Huang Shuheng, Xu Zongxuan, Chen Zhaoyue
Project Location: Taipei in Taiwan Province
Project Area: 245 m²
Major Materials: carpet, wallpaper, glass partition
Photographer: Wang Jishou
Composer: Cheng Xinchun

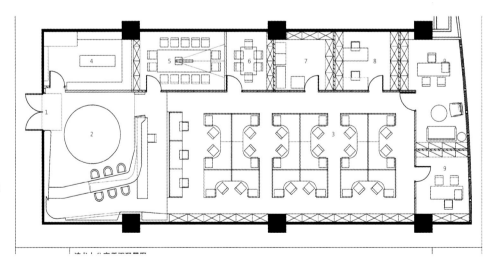

Custom Office Space

办公空间作为人们长时间工作的场所,其设计要从人的情感需求出发。合宜的空间设计能让人们之间的沟通顺畅,工作效率得以提升。一位技艺高超的设计师,同时也是一位生活大师,必然懂得生活与工作的关系,能设计出激发创意与灵感的工作空间。

就精神层面而言,如何将企业文化的精髓,在有限的时间内完整地传递给客户,则有赖设计者的纯熟技艺。办公空间作为"企业的门面",设计师必须让每个细节都与企业精神紧密贴合,使办公空间在业界独树一帜,亦使访客倍感亲切,从而达到提升品牌的目的。

在本案的设计中,设计师利用白色作为墙面与顶棚的主色调,以明亮的色系与轻盈的质感给访客的感官以舒适的体验。旋绕在空间中的白色线条,犹如白色的涡流,让访客仿佛随之旋入艺术幻境。白色是色彩之源,适宜成为办公室的统一基调,白色的背景墙面也让产品的展示效果更加出彩。

多彩多姿的织锦样品整齐地陈列着,俨然成为一个独立艺术品。大地色调的织毯铺满了洽谈空间,与墙面摆放的横向地毯互相呼应,提供给访客"脚踏实地"的真实感受。走进主要的办公区,可看见设计者选用蓝色玻璃分隔内、外空间,与门厅的大片纯白及办公区的大片铁灰相映成趣,为与客户之间的洽谈带来了轻松感,也保持了工作时应具有的严谨气质。

As a location for people's long-time work, the design of office space should start from people's emotional requirements. Appropriate space design should make the communication among people more fluent, uplifting the work efficiency. A designer with high techniques should also be a master of life. He should understand the relationship between life and work and create a work space that can inspire innovations and inspirations.

From the aspect of spiritual sphere, how to convey the essence of the corporate culture to the customers within the limited time relies on the mature skills of the designer. As the company's appearance, every detail of the office space should be closely interconnected with the corporate spirits, making the office space become peculiar, attractive to the visitors and uplifting the brand name.

For the space design, the designer applies white as the tone color for wall and ceiling, with bright color systems and light texture creating some cozy experiences for the travelers. The white lines in the space are like white vortex, inviting people inside the artistic dreamland. White is the source for colors, quite appropriate for being the integral tone of the office. White background wall makes the presentation effects of works more distinct.

The colorful brocades are displayed in order, which are just like some independent artistic objects. The woven carpets of ground color cover the negotiation space, corresponding with the horizontal carpet on the wall, providing some down-to-earth real sensations for the visitors. Upon entering the main office area, you can find that the designer selects blue glass to separate interior space from the exterior space, with the hallway grand white color and the large iron gray color of the office area bringing out the best in each other, creating some relaxation for the negotiation among customers, while maintaining the due rigid temperament during work.

设计师的办公室 | Designer's Office

设计单位：问境空间艺术工作室
设 计 师：刘建达
项目地点：福建省福州市
项目面积：120 m²
摄 影 师：周跃东

Design Company: Wenjing Space Art Studio
Designer: Liu Jianda
Project Location: Fuzhou in Fujian Province
Project Area: 120 m²
Photographer: Zhou Yuedong

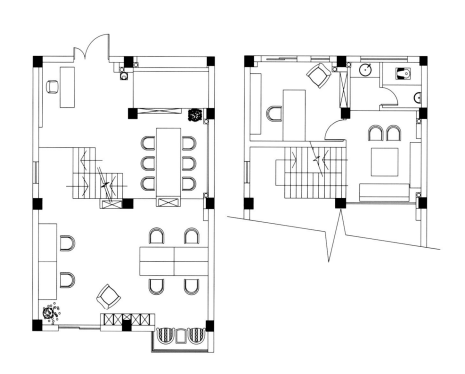

Custom Office Space

本案为一联排别墅中一、二层改造的办公空间设计，空间面积不大，却因为建筑本身的错层结构得以展现出一种既单纯又丰富的视觉体验。

设计师并没有使用太多的设计手法，也没有应用太多的复杂材料，而是通过一些材料本身所具有的肌理来突出空间本身的质感。另外，设计师对原有空间环境的优点进一步提炼，以突出空间设计的主题。

由于是临时过渡的办公室，所以在设计之初就设想如何在有限的预算中寻求一种更合理的设计方案，也为日后面对同样问题时提供可参考的解决方案。总的来说，这个方案带有一些探索的味道，没有太繁杂的创意，只想在有限的面积下，尊重原有的空间结构，营造出一个舒适的工作环境。

This project is a townhouse office space design for first and second floor regeneration. The space size is not that big but displays some pure and abundant visual experiences out of the self scattered construction of the building.

The designer does not use too many design approaches, nor much too complicated materials, highlighting the self texture of the space with the self-owned nature of the materials. Other than that, the designer makes some refinement towards the features of the original space environment, thus highlighting the design themes.

As this is a temporary office, at the start of the design, the designer tried to make out how to get some appropriate design concept with the limited budget, thus preparing references for issues in the future. Generally speaking, this project has some taste of exploration, with no complicated innovations, but respects the original space structure with the limited area and produces a cozy work environment.

万科润园办公空间   Office Space for Vanke Runyuan

设计单位：张纪中室内建筑
设 计 师：张纪中
项目地点：湖北省武汉市
项目面积：720 m²
摄　　影：牧马REN商业空间摄影机构　吴辉

Design Company: Zhang Jizhong Design Ltd.
Designer: Zhang Jizhong
Project Location: Wuhan in Hubei Province
Project Area: 720 m²
Photographer: Muma REN Business Space Photography Institution, Wu Hui

万科润园保留了原生的林木、生态系统及517工厂的工业痕迹。住宅高低错落于精心保留的原生林木间，没有抹去珍贵的岁月痕迹，而是谦逊地将新的事物巧妙注入其中，并置共生。

设计师遵循原地块的空间分隔秩序，将建筑与原生林木的尺度关系做出巧妙的协调，使生活空间延伸到外部的精神世界中。

本案为办公空间的室内设计，要表现出奢侈华丽的富贵感，同时又要避免浮夸。将悠久的历史积淀和深厚的文化底蕴表达出来，赋予空间高贵而不张扬、时尚而不妩媚的独特气质。

大厅拥有5 m的挑高空间，前台背景墙是整面的护墙板，真皮沙发、沙发椅，木格的吊顶，配以大型的铁艺吊灯，大气优雅，意境深远。家具强调简洁、明晰的线条和优雅、得体的造型。大型的落地玻璃，模糊了室内与室外的界线，真正做到了"你中有我，我中有你"。

设计师旨在打造全新的工作室环境，企业团队和个人形象通过开放的办公家具陈设和色彩设计得到了极好的展现。通过开放式的规划，设计出了温馨、友好、轻松的环境，缓解了工作中的压抑感。

Vanke Runyuan retains the original wood, ecological system and the industrial trace of 517 factory. The high and low residences are scattered among the carefully maintained original woods, with precious traits of the time, instilling inside with new matters in some modest way, attaining coexistence.

The designer follows the space separation orders of the original plot and makes some ingenious coordination towards the scale relationship between building and the original woods, making life space extend to the exterior spiritual world.

This project is the interior design for office space, displaying the magnificent sphere while avoiding showy tendency. The design displays the long-lasting historical sediments and the profound cultural connotations, entrusting the space with peculiar temperament which is noble but not showy, fashionable but not hyperbolic.

The hall has 5m of high space, the background wall for the reception wall is whole-wall wainscot wall. Real leather sofa, sofa chairs and wooden ceiling are accompanied with large iron art droplights, displaying magnificent, elegant atmosphere and far-reaching artistic conceptions. The furniture emphasizes on concise, clear lines and elegant, proper format. The grand French glass dims the separation between interior and exterior space, attaining the effect of integration.

The designer aims to create brand-new studio environment and the corporate team and individual image acquire perfect presentation through open office furniture layout and color design. The open style planning designs warm, friendly and relaxing environment, relieving the sense of repression during work.

诗意办公 | Poetic Office

设计单位：香港吉思设计有限公司
设 计 师：刘程翰
项目地点：福建省福州市
项目面积：132 m²
主要材料：麻毯、麻布、玻璃、方钢
摄 影 师：吴永长

Design Company: Hong Kong Jisi Design Co., Ltd.
Designer: Liu Chenghan
Project Location: Fuzhou in Fujian Province
Project Area: 132 m²
Major Materials: fiber carpet, linen cloth, glass, square steel
Photographer: Wu Yongchang

本案将中式元素与具有现代感的材质相融合，为室内带来轻松氛围的同时也不失办公空间所特有的庄重感，满足了高端客户的审美需求。

一进入接待空间，半透明材质上的写意荷花就向来者表达出一种宁静、悠远的气息。而朴素的石材质地的长桌，与环境搭配得十分融洽，表达出设计师崇尚自然的质朴心境。在这没有多余装饰的空间中，顶棚上方垂坠的水晶吊灯显得光彩夺目，它像一个跳跃的音符，让空间变得生动起来了。

空间的布局将私密空间和开放空间很好地区分开，但却不是完全地阻隔，形成了动中有静、静中有动的氛围。各个空间仅以古朴的置物架进行遮挡，周围布置延续了一贯的简洁风格，显示出整洁、利索的办公环境。唯有墙壁上奔腾的骏马，向客户传达出公司自强不息、勇于挑战的精神及远大抱负。

Custom Office Space

This project integrates Chinese elements with materials of modern feel, creating some relaxing atmosphere for the interior space while not lacking in the peculiar solemn feel of office space, meeting with the aesthetic requirements of high-end customers.

Upon entering the reception space, the lotus painting on the semi-transparent materials presents some tranquil and everlasting atmosphere. The primitive long table of stone texture is in high accordance with the environment, presenting the simple moods of the designer advocating nature.

The space layout finely separates the private space from the open space, but not in total separation, forming some atmosphere of dynamic in quiet and quiet in dynamic sphere. Each space is only sheltered with primitive shelf, with the surrounding space continuing the consistent concise style, presenting tidy and clear office environment. The galloping horses on the wall convey the corporate philosophy and grand expectations of constantly striving to become stronger.

BEYOND BUY INC I | BEYOND BUY INC I

设计单位：福州宽北设计机构
设 计 师：木水
项目地点：福建省福州市
项目面积：220 m²
主要材料：素水泥、乳胶漆、钢化玻璃、金属砖、不锈钢订制品、木纹白理石

Design Company: Fuzhou Comeber Design
Designer: Mu Shui
Project Location: Fuzhou in Fujian Province
Project Area: 220 m²
Major Materials: plain cement, emulsion paint, tempered glass, metal brick, stainless steel articles made to order, wood grain white marble

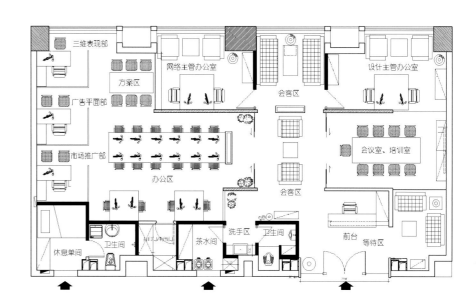

设计理念： 设计师从"简约纯粹，高利用率"的需求入手，注重材质的特性，强调功能的完善，打破办公室设计的常规布局，从家具选择到装饰等细节处入手，与空间的整体规划协调起来。

装修造价： 装修成本的控制成为本案的核心问题，硬装设计的恰到好处，为后期软装设计争取了更多的预算，后期软装陈设投入比重的提升很大程度上提升了空间的视觉品位。

材料选择： 素水泥的大量使用，使块面间的关系简洁利落，以材料本身的特性来塑造相对纯净、质朴的建筑风格。

平面布局： 本案的平面规划设计师做得非常充分，规划后 160 m² 的实际利用面积超出了对空间实际利用率的期望。

Design concept: The designer starts from the requirements of "concise, pure and high use rate," emphasizing on the specialties of the materials, completeness of functions, breaking through the common layout of office design. Starting with furniture selection and decorations and other details, the design is in accordance with the whole space planning.

Plane Layout: The designer has done a great job in the project's plane planning. The practical use area of $160\,m^2$ after planning is beyond people's expectations for the space's practical use rate.

Materials Selection: The enormous use of plain cement allows the relationship among blocks to be concise and clear, constructing a comparatively pure and primitive building style from the self features of the materials.

Decoration Cost: The cost control of decoration is a core issue for this project. The appropriate hard decoration design gets more budget for the latter stage soft decorations. The uplifting the investment proportion of latter stage soft decorations layout has greatly promoted the visual taste of the space.

东莞虎门实业集团办公室
Dongguan Humen Industrial Group Office

设计团队：KSL 设计事务所
设 计 师：林冠成、温旭武、马海泽
项目地点：广东省东莞市
项目面积：2500 m²
主要材料：灰橡木、灰木纹大理石、黑麻石、皮革、壁纸、黑钢、夹丝玻璃

Design Company: KSL Design Studio
Designers: Lin Guancheng, Wen Xuwu, Ma Huize
Project Location: Dongguan in Guangdong Province
Project Area: 2500 m²
Major Materials: gray oak, gray wood grain marble, black granite, leather, wallpaper, black steel, wired glass

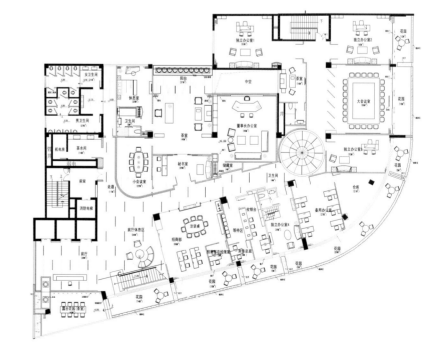

Custom Office Space

本案的核心设计要求就是希望营造出一个充满活力与激情,但又温馨的办公环境。抛却传统办公空间的冷静、肃穆的气氛,在生活与工作中找到一个平衡点,通过空间的设计将生活与工作结合起来,找到最佳的落点。

从平面布局到空间最终的氛围营造,从材料的选择到家具的布置,从灯光的营造到配饰的搭配,设计师都经过精心的考量,使每一处细节的设计力求精致而细腻。整体色调为稳重中不失温馨的褐色调,局部搭配明黄、白色,总体上给人以明快、雅致的感受。在材料上精心挑选了皮革、黑钢、木纹石、灰橡木等主材,这些材料的搭配大气中透露出典雅的基调,非常适合办公空间的设计。

The core design requirement of this project is to create a warm office space full of energy and passion. Getting rid of the quiet and solemn atmosphere of traditional office space, finding a balance between life and work, this project tries to find the perfect point through integrating life and work with the space design.

From plane layout to the final space atmosphere creation, from materials selection to setting of furniture, from lighting creation to collocation of ornaments, through meticulous considerations, the designer attains delicacy and refined quality in design of every detail. The whole color is sedate and warm brown color, accompanied with bright yellow and white to some parts, conveying some clear and elegant sensations to people in the overall sense. As for materials, the designer carefully selects materials such as leather, black steel, wood grain stone, gray oak, etc., displaying some elegant tone in the magnificent collocation, quite suitable for office space design.

深圳粤华集团办公室 Shenzhen Yuehua Group Office

设计团队：KSL 设计事务所
设　计　师：林冠成、温旭武、马海泽
项目地点：广东省深圳市
项目面积：1500 m²
主要材料：非洲胡桃木、地毯、皮革

Design Company: KSL Design Studio
Designers: Lin Guancheng, Wen Xuwu, Ma Huize
Project Location: Shenzhen in Guangdong Province
Project Area: 1500 m²
Major Materials: African walnut, carpet, leather

并不是纯粹的为了设计而设计，设计师想在商业空间与生活空间中找到设计的平衡点，重新构筑空间与人之间的关系，使现代办公空间更加人性化、生活化，符合现代人对新型办公空间的心理诉求。

在设计手法上，设计师构造了一个建筑中的建筑，笔直的线条、丰富的灯光层次，大理石、地毯、皮革与木材等各种不同质感材料的相互搭配，使空间气氛在庄重沉稳中透露出温馨的情感。使客户在洽谈生意时，还能够享受到环境带来的喜悦，为洽谈营造出了一个良好的氛围。

设计赋予空间以灵魂，使之既灵动又刚毅。而设计也能够赋予空间以情感，实现人与空间的互动与交流，只有使人在空间中体验到一种情感的依赖，才能更好地发挥人的主观能动性，创造出更加丰硕的工作成果。

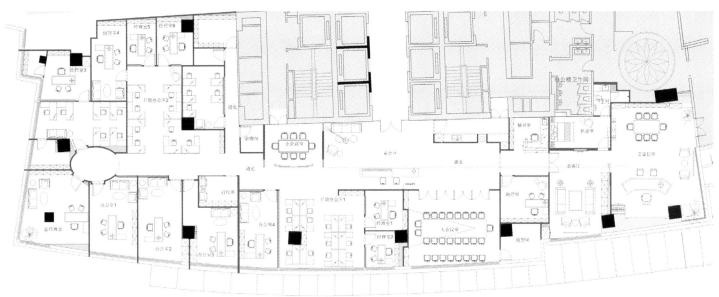

Custom Office Space 211

This is not purely a design for design's sense, the designer tries to find the balancing point for design between commercial space and life space, renewing the relationship between space and human beings, making the modern office space more human and life-like, in accordance with modern people's psychological demands for new office space.

In design approach, the designer constructs a building in building, with straight lines and abundant lights layers. The mutual collocation of different texture materials such as marble, carpet, leather and wood makes the space atmosphere send out some warm emotions in solemnity and sedateness. While the guests are negotiating business, they can have the pleasure brought by the environment, surrounded by this excellent atmosphere.

The design entrusts the space with souls, making the space dynamic, resolute and steadfast. The design can also entrust the space with emotions, realizing the interaction and communication between human beings and space. Only when people experience some emotional reliance in space, can they better display the human being's subjective initiative, creating much plentiful and substantial work results.

Shude Headquarters Office

树德办公总部

设计单位：广州道胜装饰设计有限公司
设 计 师：何永明、崔华峰
项目地点：广东省广州市
项目面积：1500 m²
摄 影 师：罗广

Design Company: Guangzhou Daosheng Design Co., Ltd.
Designer: He Yongming, Cui Huafeng
Project Location: Guangzhou in Guangdong Province
Project Area: 1500 m²
Photographer: Luo Guang

本案的设计遵循现代空间的秩序感，同时追寻古典主义的气质。设计师运用横竖相交的线条来体现空间的秩序感，同时也使得空间有了延伸性，引导人们进入其他空间。颠覆了传统办公空间过于平实的既定印象。

在立面的造型分割与材质运用上，设计师借用装饰艺术风格中平行线规则排列的造型方式，大量使用鸡翅木与三聚氰胺板。藉由虚实相映的格栅与玻璃，将自然光或是艺术光源隐藏起来，赋予了空间无穷的趣味。

Custom Office Space

The design of this project follows the order feel of modern space, while seeking the temperament of classicism. The designer displays the order feel of the space with lines of horizontal and vertical intersections, providing the space with extensive power, while guiding people into other spaces, overturning the established plain impression of traditional office space.

As for facade division and materials application, the designer applies the format of parallel line in Art Deco, making a lot use of door frame veneer and melamine board. With the grating and glass of true and false reflection, natural light and artistic light source are hid in a subtle way, entrusting the space with limitless interests.

捷致办公室 | Jiezhi Office

设计单位：玄武设计
设 计 师：黄书恒、许棕宣、董仲梅
项目面积：300 m²
主要材料：壁纸、方块毯、喷漆
摄 影 师：王基守
撰　　文：程歆淳

Design Company: Sherwood Design Group
Designers: Huang Shuheng, Xu Zongxuan, Dong Zhongmei
Project Area: 300 m²
Major Materials: wallpaper, square block carpet, paint
Photographer: Wang Jishou
Composer: Cheng Xinchun

"捷致科技"办公室的设计案,在设计中将企业精神作为设计主轴,空间中也隐藏了设计者的幽默。

"捷致科技"主要生产计算机游戏的外围产品,游戏产业的魔幻炫目及行业的日新月异,都决定了其办公空间的时尚感。因此,设计者以深沉的黑色打底,用大面积的黑色墙面营造出变幻莫测的感受,仿若一口深不见底的黑井,神秘的回音在空间中盘旋回荡,激发着访客继续探索。黑色,也隐喻着游戏产业的未来,就像等待开启的计算机屏幕,将抽象的企业精神具象显现,成为办公空间的设计主旨。

除了充盈着的黑色之外,设计者采用明度不一的蓝色作为第二色彩,勾勒空间的亮点。LOGO背后透射出的蓝光,在幽暗的空间中隐隐浮动,好似计算机屏幕上不停闪烁的光标,引导人们的视觉与想象力。浅蓝色勾勒出英文字母与企业标志,在点缀空间色调的同时,也为以白色为基底的工作空间揭开了序幕。

仔细观察柜台的背景墙,访客必须站在适当的位置,以最准确的角度,才能清楚地看到隐藏在上面的黑色曲线,随着空间的韵律而起伏。幽黑的线条呈现着二维与三维向度的思辨。同时,访客也能从纵横起伏的纹路中理解设计者的用心。

For the design of this project, it has the corporate spirit as the design axis, while hiding the humor of the designer.

"Jiezhi Technologies" mainly produces the actual products for computer games, with the dazzling computer games industry and the progress with each passing day determining the fashion feel of the office space. Thus, the designer has profound black color as the base and applies large area of black wall to produce the unpredictable sensations, just like a bottomless black well. The mysterious sound is hovering and echoing in the space, inspiring the visitors to proceed with the exploration. Black implies the future of games industry, like the computer screen waiting to be turned on, displaying the corporate spirit in a concrete way, becoming the design tenet for the office.

Apart from the abundant black colors, the designer applies blue of varying brightness as the second color, delineating the highlight of the space. Blue color transmits from behind the LOGO, dimly floating in the dark space, just like the glittering cursor on the computer screen, guiding people's visual sight and imaginations. Light blue color draws the outline of English letters and the corporate logo, dotting the space color tone, while producing a prelude for the work space with white color as the base.

In order to carefully observe the background wall of the counter, the visitor needs to stand in some appropriate location and clearly sees the black curves hidden on it which fluctuate with the space rhythms with the most exact angle. The black lines display the philosophical thinking of two dimensional and three dimensional aspects, while the visitor can perceive the meaning of the designer from the fluctuating patterns.

禾大办公区设计 | CRODA OFFICE

设计单位：usoarquitectura / Gabriel Salazar y Fernando Castañón
设 计 师：奥克塔维奥·瓦斯克斯
施工单位：Atxk
家具设计：Knoll
项目地点：墨西哥联邦区
摄　　影：Héctor Armando Herrera

Design Company: usoarquitectura / Gabriel Salazar y Fernando Castañón
Designer: Octavio Vásquez
Constructor Company: Atxk
Furniture Company: Knoll
Project Location: México D.F.
Photography: Héctor Armando Herrera

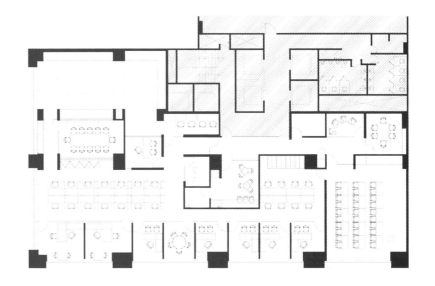

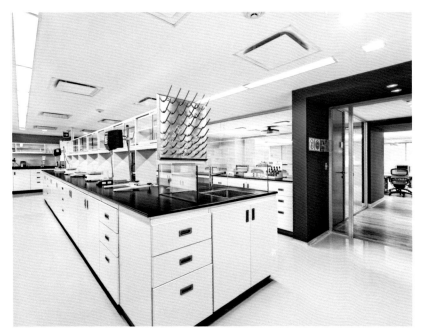

禾大公司是一家专业从事用天然材料打造化工产品的国际化公司。其办公区改造是一个很有意思的项目。知名设计师与公司通力合作，不仅要打造出独特的办公空间，还要改变公司同仁的工作方式。该项目要通过一系列重要的改造以赋予原有的空间以新的环境特色。

要在室内空间中进行一些改变，很重要的一点是打造出开放式的空间，尤其是将新形式的工作动态融入整个空间架构中。这样的改造方案面临的一大挑战是，将原先封闭式的办公区改造为开放式的，这一点主要是通过打造板凳式家具系统实现的，这也进一步提升了团队协作精神。

该项目还涉及一处实验室，该实验室对整个公司的日常运作非常重要。主要在其中开展一些分析、展示、培训等活动，这些对于客户服务来讲都是至关重要的组成部分。该实验室成为整个项目的核心区，并因其独特的设计而独具魅力。当人们身处工作区时，可以看到这处风格独特的实验室。

培训室可容纳30人，与入口大厅、会议室、咖啡厅融为一体。禾大员工和访客均可使用这处咖啡厅。会议室的外观、大小各不相同，以满足公司运营不同情况下的需求。

The project for the redesign of Croda offices – an international company specialized in the production of chemical products with natural origins – was an interesting process in which architects Gabriel Salazar and Fernando Castañón were involved with the client to change, not only the space, but the way they work. Important modifications had to be done in the same space to generate a new environment.

To create changes from the interior it is very important have an open communication channel all the time, especially to transmit the new work trends to the entire organization. One of the main problems was changing the offices from closed to open space, which was solved by implementing bench furniture systems to increase the team work and limited filing space per person to use less paper.

The program included a laboratory – which occupies the fifth part of the area – very important for the activities of the company. This laboratory is for analyzes and demonstrations and training, both activities are essential part to the customer service. The laboratory was not considered to be a protagonist space but the axis of productivity in an unusual and very attractive way because it can be seen from the operative area.

There also included a training room for 30 people, which merged with the entrance hall, meeting rooms, and cafeteria are that are used by both visitors and CRODA staff.

229

Paga Todo 办公区设计
Paga Todo Office

设 计 师：Gabriel Salazar y Fernando Castañón
项目面积：2000 m²
主要材料：纸面石膏板、玻璃、预制木质材料、方块地毯

Designers: Gabriel Salazar y Fernando Castañón
Project Area: 2000 m²
Main Materials: drywall, glass, millwork, carpet tiles

对于一家公司的室内设计和具体运作而言，所具有的空间是主要决定因素。针对 Paga Todo 新建的办公区制定室内设计方案，对于设计团队而言是个不小的挑战，既要满足客户要求，又要在购物中心打造一处 2000 m² 的办公空间。

这处大型的木质箱体结构对周边空间设计给予了充分尊重，并欢迎着公司合作伙伴和访客的到来。箱体内部设置了接待区、服务区和会客厅。箱体顶部为一处为经销商服务的个性化空间，拥有欣赏周边金融区的全景式视野。

客户希望能在办公区内打造一处休息室风格的咖啡厅，类似于酒店大堂。因为在新办公区建成之前，公司合作伙伴都特别喜欢在附近的咖啡厅里会面谈工作，可以享有非常闲适的空间氛围。该空间的服务设施非常齐全，对于访客而言是一大惊喜，因为这里配备了免费上网电脑、点心、饮料等。

基于客户要求,空间色调较为肃穆,没有使用夸张的色彩。白色、褐色和绿色相互辉映,搭配着橡木制家具和木制品。设置了会客厅的三个部分将整个空间分隔开来,同时在各个工作单元之间形成一定的过渡空间。

为了引入尽可能多的自然光,建筑立面上设置了很多大型的垂直式条状结构。而大多数的墙体又不与顶棚相接,这种设计方案是为了将各种不同的自然光源充分利用起来。

Space is the main factor that determines the interior design and operation of a company. The new corporate offices for Paga Todo presented a particular challenge because it was necessary to adapt to the clients demands and a 2,000 sq m area in a shopping center.

A big wood box, inserted respecting the surrounding design, greets everyday collaborators and visitors. Inside the box were located the reception, support area and interview halls, on top of it –with a panoramic view of the finance area- the personalized area to serve the dealers.

The client decided to implement in their office a lounge style cafeteria –like a hotel lobby- because before the relocation the majority of the collaborators preferred to work and meet in the close by cafeterias to enjoy a more relaxed ambiance. This space has all the necessary services and it is a nice surprise for the visitors because there are screen, complimentary computers with Internet, snacks and drinks.

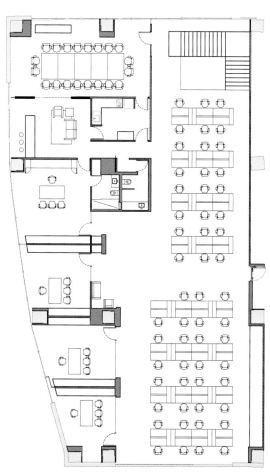

The color palette –asked by the client- is very sober and with no risk. White, beige shades with accents in a dry green and the oak of the furniture and woodwork. Three sections with meeting halls divide the space generating references and transitions between the work cells.

For natural light big vertical stripes were open on the façade of the shopping center and most of the walls were not built ceiling height to make the most of the different natural light sources of the building.

墨西哥城某办公空间 | Corpovo Ifahto

设计单位：ARCO Arquitectura Contemporánea
Arq. José Lew Kirsch
Arq. Bernardo Lew Kirsch
项目面积：640 m²
主要材料：环氧树脂、大理石、复合地板
摄 影 师：杰米·纳瓦罗

Design Company : ARCO Arquitectura Contemporánea
Arq. José Lew Kirsch
Arq. Bernardo Lew Kirsch
Project Area: 640 m²
Major Materials: epoxy resin, marble, laminate flooring
Photography: Jaime Navarro

Custom Office Space

IFAHTO新建办公区设计的主要考量因素囊括了团队合作、平衡性及灵活性。设计团队的主要目标是以创新性的方式将为客户提供的优异服务展现在人们眼前。人们一进办公区入口，即可清晰看到公司的主要特色和核心理念。

整个项目选用的色调非常简洁，以很多中性色彩，诸如白色、黑色和不同色调的灰色为基调。色彩的选择使整个空间具有极高的辨识度。而设置在开放区域的诸多元素，诸如绿色墙体、禅意花园、装饰性挂画、黑色板材及倾斜的有色墙体等，都很好地融入空间内，使该办公区显得不拘一格、独具特色。

办公区内的各处主要空间都被很好地整合在了一起，诸如行政区、工作区、服务区、公共区和人行通道等。各个经理办公室所处位置也进行了巧妙设置，以便各位经理对其所属工作区都能一览无余。设计的主要原则贯穿整个项目之中，并保证了适当的私密性。

该办公区室内设计的一大亮点是会议室。会议室为一处悬浮式的箱体结构，比其余办公空间略高。该箱体结构外部覆以一系列的滑动墙体，墙体又以各色图像装饰，不仅体现移动效果，还确保了会议室的私密性。

创新性是该办公区室内设计的核心原则之一，为了进一步凸显这一特色，设计师选用了一些轻盈而又富于灵活性的室内家具。家具均满足了办公活动的需求，并赋予空间以无穷活力。

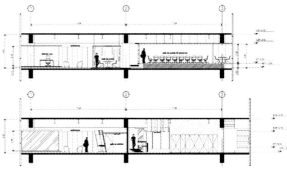

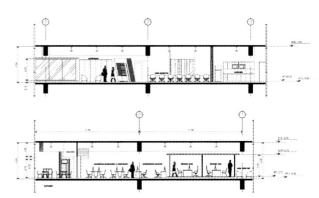

Team work, balance and flexibility were the elements considered in the design for the new offices of the agency IFAHTO. Their mission is to bring their clients excellent service in a creative way. What the agency is and believes is clearly noticed since crossing the main entrance.

The color palette selected for the whole project is plain and neutral colors such as white, black and a wide variety of gray shades are predominant. The color selection allowed us to define very well the spaces, and in the different open areas elements such as: a green wall, Zen garden, decorative graphics, blackboards and a tilted color wall were incorporated to give the agency its own personality.

The zone marking studies helped define the principle areas: direction offices, operative areas, services, common areas and circulations. The director's offices were located in a strategic way, because each one monitors the activities of their own area. The design guidelines were kept throughout the project, the only difference in this area is the increase of privacy derivative of the tasks.

Another main feature in the interior of this agency is the meeting room. It is a floating box located higher than the rest of the areas. The box is covered with a series of slide walls –that look like scales- covered with graphic images that provide movement and privacy at the same time.

Innovation is one of the main axes of the agency and what defines its lifestyle in order to stand it out light and flexible furniture was selected. It adapts to the needs and dynamic of the activities involving the user in the creation of a space in constant change.

BEYOND BUY INC II

BEYOND BUY INC II

设计单位：福州宽北设计机构
设 计 师：木水
项目地点：福建省福州市
项目面积：90 m²
主要材料：素水泥、金刚板、乱石板岩、罗非岩、墙基布、芬琳乳胶漆

Design Company: Fuzhou Comeber Design
Designer: Mu Shui
Project Location: Fuzhou in Fujian Province
Project Area: 90 m²
Major Materials: plain cement, laminate flooring, slate, rock, wall cloth, emulsion paint

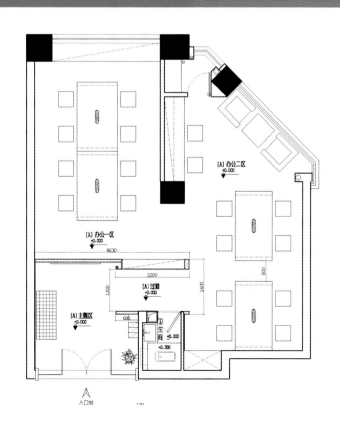

格调：BBI 公司的第二处办公场地依然延续黑、白、灰的素描风格。通过对空间主次关系精致的刻画，使得空间呈现愉悦、明朗、轻快、幽静的视觉观感。黑、白、灰的渐变可以衍生出无穷尽的空间序列及视觉变化。由于黑、白、灰没有色彩，它的美便寄于形。空间的大小、疏密、高低排序带出的节奏及次序感，都将对空间产生绝对的影响。

综述：本案的设计风格简练通畅，空间利用率的提高显示了设计师深厚的设计功底，空间色调平衡、和谐。空间陈设不喧宾夺主，而是恰到好处。

材料：本案的地面使用素水泥找平，并做环氧树脂透明漆涂层。主题区的墙面用乱石堆砌后扫白，以平衡墙面的厚重及凌乱感。所有家具均现场制作并饰以白色烤漆，U 形的出入口设置了隔断及资料柜，与卫生间的隐形门一起采用金刚木地板，自然实现了无缝拼接。在两块主题墙面上都拼贴了墙基布，还在局部采用灰色乳胶漆及罗非岩涂料饰墙。

规划：空间整体呈钻石形，其实际利用面积仅为 68 m²。功能空间分为主题接待区、18 人的公共办公区、卫生间、茶水室，其对空间利用率的要求较高。

Style: The second office space for BBI corporation also continues the black, white and gray sketch style. Through delicate sketching of the space's primary and secondary relationships, the space displays pleasing, bright, brisk and tranquil visual expressions. The gradual changes of black, white and gray colors can produce limitless space orders and visual variations. As these three colors do not have bright appearance, their beauty is displayed in the format. The scale, density, high and low orders of the space would create rhythms and order feel which would have some absolute influence on the space.

Planning: The space has diamond format, with practical use area of only 68 m². The functional space is divided into theme reception area, public office space for 18 people, wash room and tea room which have high requirement for the use rate of the space.

Materials: The ground of this project has plain cement, with transparent coating of epoxy resin. The wall of the theme area is decorated with scattered rocks, balancing the heaviness and messy feel of the wall surface. All furniture is made on site and decorated with white stoving varnish. There is a partition and material cabinets at the exit and entrance of U format, attaining seamless connection with the invisible door of the wash room through laminate flooring. Wall cloth is applied on both theme walls, decorated on some parts with gray emulsion paint and rock coating.

Summary: The design style of this project is concise and fluent, with high space use rate demonstrating the profound design capabilities of the designer, displaying balanced and harmonious space color tones. The space layout is not showy at all, but with appropriateness.

智威汤逊北京公司（JWT）
J.Walter Thompson Company in Beijing

设计公司：金碧室内设计工程有限公司
设 计 师：郭建斌
项目地点：北京
项目面积：1800 m²
主要材料：水泥、乳胶漆、石膏板、地毯
摄 影 师：陈婧

Design Company: INFINITY DESIGN & ENGINEERING
Designer: Guo Jianbin
Project Location: Beijing
Project Area: 1800 m²
Major Materials: cement, emulsion paint, plasterboard, carpet
Photographer: Chen Jing

本案的设计难点是在有限空间内容纳200个员工，不但要具备办公的基本功能，其设计还要舒适、严谨、突出行业的特点。

JWT是一个充满创意的国际团队，拥有极高智商的人对于空间尤为敏感，所以人和空间的对话就从盒子的形态开始演绎。每个人每天的生活的和工作都离不开盒子，人们在盒子里扮演不同的角色。盒子的块面组合也是自然界最简单、最美好的状态，块与面的结合、体与块的关系、面与面的结合无不涵盖JWT的企业文化内涵，因此，本案获得了业主极高的肯定。

材料选用上考虑经济性和节能减排的要求，常见的水泥地面和新加坡地毯相互对比，让使用者既轻松自然，又精神抖擞。白色和绿色是整个空间的主色，色彩斑斓的家具和饰品成为了空间主体。

设计师在仔细研究了空间的结构特点和业主的需求后，采取了一种既有组织又自由松散的布局方式。各个体块进行有机的穿插，将所

Custom Office Space

有的入口弧形放大，给人一种视觉上的舒适感。将公司分成三大组织结构，三大组织用树叶图案的地毯连接起来，一眼望去既自然又有生命感，从而创建整个空间的流动性。可以让来访者自然而然地从一端走到另一端，而感觉空间很大，也自然地铺陈了空间的交通动线。

The challenge for the design is to accommodate 200 staff in the limited space, possessing not only fundamental functions of office, but also cozy and rigid design highlighting the industrial features.

After carefully analyzing the space structures and requirements of the property owners, the designer applies some orderly, free and loose layout approach. There is organic alternation among different blocks, expanding all the entrance arches, creating some visual comfort. The corporation is divided into three organization structures, which are interconnected with leaf graphic carpet, showing some natural and dynamic sphere and producing the fluent nature of the whole space. The visitor can walk quite naturally from one end to the other end, sensing the grand nature of the space, while producing the transportation lines of the space.

As for materials selections, the designer focuses on economic and energy-conservation, emission-reduction requirements, with contrast of common cement floor and Singapore carpet, making the people feel relaxed and be in high spirits. White and green are the tone colors of the whole space, and the colorful furniture and ornaments become the main part of the space.

JWT is an international team full of innovative ideas and people of super high intelligence quotient are quite sensitive to the space, thus the dialogue between people and space starts from the deduction of the format of the box. Everyone's daily life and work can not

get away from the box, and people play different roles in the box. The box's block surface combinations are the simplest and most beautiful status in the grand nature. The combination of block and surface, the relationship between body mass and block, and the combination of surfaces all cover the cultural connotations of JWT. Thus, this project won high appraise from the property owners.

科大永合投资有限公司　Keda Yonghe Investment Co., Ltd.

设计单位：福建国广一叶建筑装饰设计工程有限公司
方案审定：叶斌
设 计 师：何华武
项目地点：福建省福州市
项目面积：500 m²
主要材料：地毯、金刚板、玻化砖、乳胶漆

Design Company: Fujian Guoguangyiye Architectural Decoration and Design Engineering Co., Ltd.
Project Examiner: Ye Bin
Designer: He Huawu
Project Location: Fuzhou in Fujian Province
Project Area: 500 m²
Major Materials: carpet, laminate flooring, vetrification tile, stainless steel

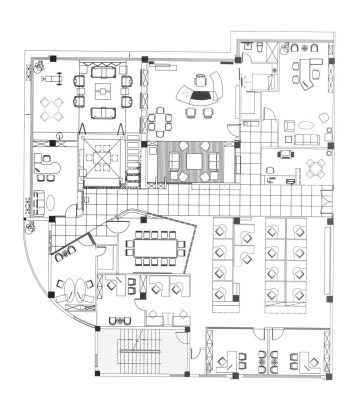

Custom Office Space

本案在设计中,以低碳、节能为设计的首要考虑,设计师大量运用低成本的物料来填充空间。黑、白元素的搭配,使整个设计充满着浓厚的现代情调。另外,设计师注重人性化的设计,合理规划平面布局,使功能分区合理,交通动线流畅,并巧妙运用几何形对墙面进行装饰,极为符合现代风格办公空间的要求。

高层管理人员的办公室布置在公司的两侧,室内布置有高端的家具设计,挑选了舒适性及现代感较强的适合长期办公的家具。

设计始终力求从整体到局部的布局,为了方便公司的统一管理,在开放性办公空间采用"一"字形布局,既宽敞又便捷,方便员工与上司之间的交流与合作。长长的日光灯管为办公提供充足的光照度。简洁的会议室墙面运用块面的处理手段,利用玻璃与墙面的分割,不仅使过道空间感更强,也使会议室通透感更强,改变了传统会议室的沉闷感。

设计师对办公空间做了横向分隔,用简洁的色彩将地面与墙面衔接在一起,结合现代化的设计手法来营造办公空间带给人们的亲切感。为了给公司营造温馨的气氛,设计师通过异形墙面的设计增强过道空间的延伸感。走廊尽头的圆形窗户象征清晨初升的太阳,圆形的吊挂与太阳联系在一起,好比公司与客户紧密联系。

In design, this project has as initial considerations of low-carbon and energy-conservation design. The designer makes a lot use of low-cost materials to fill the space. The collocation of white and black elements allow the whole design to be filled with dense modern artistic conceptions. Other than that, the designer focuses on human design, appropriately planning the plane layout, with proper functional area separation and fluent transportation moving lines. The designer makes ingenious use of geometric style to decorate the wall, quite in accordance with the requirements of modern style office space.

The offices for senior management personnels are set on both sides, with high-end furniture design inside, selecting comfortable and modern furniture suitable for long-term office requirements.

The design always pursues the layout from whole to parts. In order for the consistent management of the company, there is this linear layout in the open office space, expansive and convenient, quite suitable for communication and cooperation between staff and the managers. The long fluorescent light tubes provides some sufficient lights for the office space. The concise conference rooms' wall applies block treatment. The separation of glass and wall surface not only strengthens the space feel of the corridor, but also makes the conference rooms appear more transparent, changing the heavy feel of traditional conference rooms.

The designer carries out horizontal separation towards the office space, connecting the ground with the wall through concise colors, creating some intimate feel for people of the office space through modern design approaches. In order to create some warm atmosphere for the company, the designer strengthens the extension feel of the corridor though design of irregular shape of wall surfaces. The round window at the end of the corridor represents the rising sun in the morning, with round hanging connected with sun, like the intimate relationship between company and customers.

天安国际大厦商务楼
Office Building for Tian'an International Tower

设计单位：深圳市尚邦装饰设计工程有限公司
项目地点：广东省深圳市
项目面积：1100 m²

Design Company: Shenzhen S&D Design Co., Ltd.
Project Location: Shenzhen in Guangdong Province
Project Area: 1100 m²

接待大堂是展示公司形象的地方，本案中的大堂设计展示出专业的商务气息，给到访者以实力、专业、气派的空间印象。

电梯厅中，具有时尚色调的休闲等候区，呈现出与众不同的商务气息。办公区的通道，具有不断变化的灵活性，满足了不同办公人员不断变化的业务需求，使空间更具人性化，体现出一种新型的办公理念。茶水室的陈设布局与灯光营造使得空间格外有趣，是一处兼具休闲与娱乐的放松空间。

设计师将现代办公的理念融入设计中，使一切显得更加灵活、便捷，容纳性更强，过道墙面上的存储柜不仅打破了视觉上的沉闷感，还更加充分利用了空间。

设计师将现代办公的理念充分与空间环境相结合，创造出了兼具时尚与空间气质的完美办公环境。充足的日光引进来，使工作环境更加的开阔，明快的色调使人有一种愉悦感。

Custom Office Space

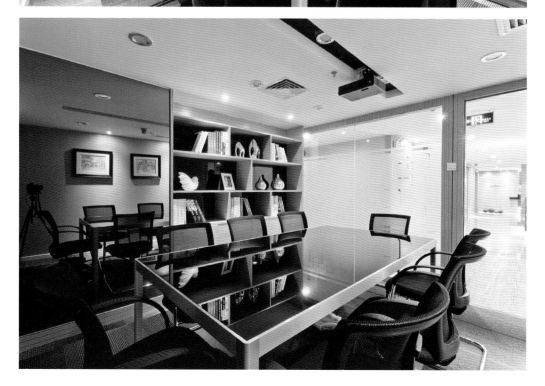

Reception hall is a place to display the corporate image. The lobby design of this project presents professional commercial atmosphere, creating some solid, professional and grand space impression to the visitors.

Inside the elevator hall, the leisure reception room of fashionable tone color displays some uncommon business atmosphere. The corridor of the office area possesses some varying dynamic nature, meeting with the varying business requirements of different office personnels, making the space more human and presenting some novel office concept. The layout of the tea room and lighting makes the space quite spectacular, as a relaxing space possessing relaxation and entertainment functions.

The designer integrates the office concept inside the design, making all dynamic and convenient, of powerful inclusive nature. The storage cabinets on the corridor wall not only break through the dumb feel visually, but also make full use of the space in a better way.

The designer fully combines modern office concept with the space environment, creating a perfect office environment with fashion and space temperament. The sufficient sunlight makes the work environment appear broader and the bright colors produce some relaxing pleasant feelings.

凯发贸易办公空间 | Office Space for Kaifa Trade Co.

设计公司：杭州观堂室内设计有限公司
设 计 师：张健
项目地点：浙江省杭州市
项目面积：750 m²
摄 影 师：王飞

Design Company: Guan Interior Design Co., Ltd. in Hangzhou
Designer: Zhang Jian
Project Location: Hangzhou in Zhejiang Province
Project Area: 750 m²
Photographer: Wang Fei

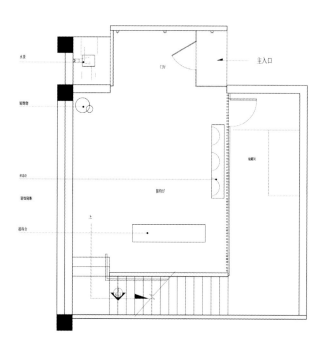

本案设计师运用现代的设计手法进行设计,在空间布局设计上层次分明,简洁的线条延伸过渡到各个办公空间,这种利落的线条节奏展现出独特的形式美感。另外,室内设计注重与外界空间的沟通,将优美的自然风光引入到室内空间,再配以恰到好处的灯光,营造出了一个简洁明朗、自然通透的现代办公环境。

办公空间的设计给人以安静、平和、整洁的整体印象。平面布置得十分规整,家具样式与整体色彩也非常统一。透过大面积的玻璃可以饱览远处的美景,这样的设计为员工营造出了一个温馨、内敛、包容、宽松的工作环境。

会所的接待处设计简单、大气,开阔的空间配合沉稳的色调,使空间的质感凸显出来。再往里走是洽谈区和红酒展示区,空间的装饰采用了现代风格的家具及饰物,加入了明快亮丽的色调,顿时给人一种愉悦、轻松的感觉。

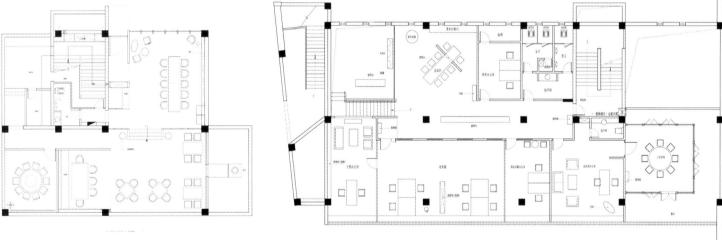

The designer applies modern design approach to carry out the design, with clear layers in the space layout design, with concise lines extending to various office spaces. This kind of brisk lines rhythms displays spectacular format aesthetic beauty. Other than that, the interior design emphasizes on the communication with the exterior space, introducing beautiful natural landscape inside the space, accompanied with appropriate lighting, producing a modern office space of concise, natural and transparent style.

The design of the club reception space is simple and gorgeous, highlighting the space texture with the combination of broad space and sedate color tone. Further inside is the negotiation area and red wine presentation area, with the space decorations applying furniture and ornaments of modern style, presenting some pleasing and relaxing sensations with the bright and brisk colors.

The office space design produces some quiet, peaceful and tidy whole impression. The plane layout is quite orderly, with unification of furniture format and whole colors. Through the large glass, one can enjoy the distant nice views, with such design producing a warm, introvert, inclusive and relaxing work environment for the staff.

凯明迪律师事务所办公室设计
Office Design for Chiomenti Studio Legale

设计公司：Stefano Tordiglione Design Ltd.
设计总监：Stefano Tordiglione
项目地点：香港
项目面积：300 m²
摄 影 师：Tanz Baig

Design Company: Stefano Tordiglione Design Ltd.
Design Director: Stefano Tordiglione
Project Location: Hong Kong
Project Area: 300 m²
Photographer: Tanz Baig

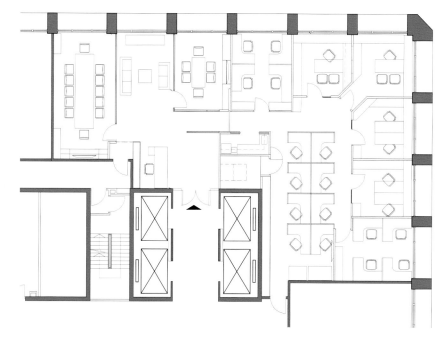

意大利风情尽显，东西风韵的全新演绎

在色彩设计上，设计师大量地应用了白色和中性色，与门后和会议室的玻璃墙相互融合，大大增强了空间感。深褐色的古典木地板令空间温暖起来，展现了空间的独特个性。

凯明迪律师事务所的办公室位于香港中环置地广场，室内的主题设计元素是简洁的线条，弧线的点缀加强了装饰效果。两者的含蓄对比表明了中国文化与意大利文化的融合与包容，是艺术与哲学的共融。不同材料如玻璃、木材、金属相互协调，散发出不一样的空间表情，展现出低调而奢华的空间气质。高质量的木材、几何圆形的柜子、不锈钢板和可移动的储物柜，都展现出设计团队对办公细节的注重。

办公室内也汇集了不少高级灯具，包括意大利的 ARTEMIDE 和瑞典的 FLUX。圆形的吊灯主要用于接待处和走廊。会议室中别致的天花造型，透射出柔和的光线，使得精致的顶棚造型好像漂浮在空中。富有现代感的书柜与各种金属、木材有机地结合起来，散发出简约宁静的感觉。

本案的设计开创了意式设计风格的先河，为东方和西方美学的融合带来了新的气息。使现代化的办公场所设计得到了全新的演绎。

Custom Office Space

Full of Italian Expressions, New Presentation of East and West Charms
The design of this project starts Italian design style, bringing new atmosphere for the integration of east and west aesthetics, creating brand-new interpretation for modern office space.

This office project is located in The Landmark, Hong Kong, with theme interior design elements of concise lines which strengthen the decorative effects. The implicit contrast displays the integration and containment of Chinese and Italian culture, as the integration of art and philosophy. Different materials, such as glass, wood and metal, are consistent with each other, sending out some different space expression, while presenting low-key and luxurious space temperament. High-quality wood, geometric round cabinet, stainless steel plate and movable storage cabinet all present the emphasis of the design team on the office details.

The office space also integrates quite a number of high-level lighting objects, such as Italian ARTEMIDE and Swedish FLUX. Round drop lights are mainly used in the reception area and the corridor. The unconventional ceiling format in the conference room sends out some soft lights, making the delicate ceiling seem to float in the air. The modern bookcase is combined in an organic way with various kinds of metal and wood, emitting some concise and tranquil sensations.

As for color design, the designer makes a lot use of white and neutral colors, integrating with the glass wall behind the door and in the conference room, greatly strengthening the space feel. Dark brown classical wood flooring makes the space appear quite warm, displaying the peculiar characteristics of the space.

IEA 总部办公室

Office for IEA Headquarters

设计单位：武汉艾亿威装饰设计顾问有限公司
设 计 师：王治
项目地点：湖北省武汉市
项目面积：1000 m²
主要材料：工字钢、水曲柳木饰面、黑色及白色混油、黑色地毯、镜面、原木色实木地板、海吉布

Design Company: Wuhan IEA Interior Design Consultants Co., Ltd.
Designer: Wang Zhi
Project Location: Wuhan in Hubei Province
Project Area: 1000 m²
Major Materials: joist steel, ash-tree wood veneer, black and white mix oil, black carpet, mirror surface, burly-wood solid wood floor, Haijibu wall cloth

本案位于武汉内环核心的LOFT SOHO，整体的玻璃幕墙可以使人们自由地北望长江，4m高的挑高空间呼应黑、白主调，强调出高效明快的办公氛围。整体木质书架铺满墙面，钢结构的小平台与大厅互动，错落有致的灯光与饰物点缀其间，使空间大气而充满趣味。室内运用了几何形式的线条来划分空间，并配合机械感十足的灯具装饰，营造出简约而时尚的LOFT办公氛围。

设计师采用现代简约的设计手法，配合对比强烈的色调，将有序的节奏感与朴素的环境巧妙融合，带来了一种介于宁静与粗犷间的享受。顶棚照明与墙面设计相配合，将室内光线均匀分布，提高了室内的光照水平。而不经修饰的顶棚、简单的色调、厚重的墙体，带来了工业文明时代的沧桑韵味，传达出空间的文化内涵。

Custom Office Space

This project is located inside the LOFT SOHO of Wuhan's inner core, with whole glass curtain wall which can allow people to watch the Changjiang River freely. 4m high space echoes the black and white tone, emphasizing on the high-efficient and brisk office atmosphere. The whole wooden bookcase covers the whole wall, with the steel structure terrace interacting with the hall, dotted with scattered lights and ornaments, making the space grand and full of interests. The interior space applies geometric lines to separate the space, accompanied with lighting accessories of machine feel, creating concise and fashionable LOFT office atmosphere.

The designer applies modern concise design approach, accompanied with color tones of intensive contrast, ingeniously integrating orderly rhythmic feel with plain environment, and bringing some enjoyment between tranquility and roughness. The ceiling lighting is combined with wall design, distributing interior light evenly and highlighting the interior lighting quality. The crude ceiling, simple color and profound wall convey the lasting appeal of vicissitudes in industrialized civilization, displaying the cultural connotations of the space.

杨大明办公室 | Yang Daming's Office

设计单位：杨大明设计顾问事务所
设　计　师：杨大明、吴金云
项目地点：湖北省武汉市
项目面积：700 m²
主要材料：素水泥、竹面板、水磨石、夹胶玻璃、灰镜
摄　　影：牧马REN商业空间摄影机构　吴辉

Design Company: Yang Daming Design Consultant Firm
Designer: Yang Daming , Wu Jinyun
Project Location : Wuhan in Hubei Province
Project Area : 700 m²
Major Materials: plain cement, bamboo panel, terrazzo, laminated glass, gray mirror
Photography: Muma REN Business Space Photography Institution, Wu Hui

设计师在平面规划中遵循"实用性、功能化与人性化管理充分结合"的原则,既考虑办公需求与工作流程,也考虑员工之间的相互交流,以及功能区域之间、工作与休闲区之间的相互协调,科学合理地划分功能区域。经过合理的布局与规划,在满足各种办公需求的同时,塑造出简洁、自然、舒适的工作环境,充分体现出企业的形象感与文化内涵。

宽敞的二层空间中,平面分布以入口为界线,划分出不同的功能区域,并按照人体工程学的原理选择合适的办公家具。整个分区既能满足各部门的功能要求,也实现了人员的有效分流,同时又使得各个部门之间的沟通比较顺畅。

空间以素水泥为主要材料来铺陈，与木饰面完美搭配起来，呈现出富有质感的有机形态，体现出现代、质朴、自然的设计理念，彰显出实用、环保与人性化相结合的气度。

入口处的生态竹、鸟笼的造型别具一格，与使用了绿色玻璃的楼梯扶手、二楼围栏遥相呼应，使空间中顿时充满了生机与活力，也为员工在忙碌的工作之余营造出了一个放松心情的氛围。

In the plane planning, the designer follows the principle of utmost connection of practical, functional and human management, not only considering office requirements and work processes, but also considering the mutual communication among staff and the coordination of functional area, work and leisure zone, dividing the functional areas in a scientific and appropriate way. Through proper layout and planning, while meeting with various office requirements, the design produces a concise, natural and comfortable work environment, fully representing the image feel and cultural connotations of the corporation.

For the expansive second floor space, the plane layout has the entrance as the boundary line, creating different functional areas, and selecting proper office furniture based on the principle of ergonomics. The whole space division not only meets with the functional requirements of various sections, but also attains efficient stream distribution, while making the communication among different departments more smooth.

The space has plain cement as the main material to decorate the space, in perfect collocation with the wood veneer, displaying organic format of great texture, manifesting modern, primitive and natural design concept and representing the atmosphere of connection of practicality, environmental protection and humanization.

The format of the ecological bamboo and bird cage at the entrance displays unique format, echoing the staircase handrail and the fencing on the second floor of green glass, making the space be filled with energetic and dynamic atmosphere, while providing the staff with some relaxing environment away from the busy work.

云裳·婚纱工作室 / Yunshang·Wedding Dress Studio

设 计 公 司：深圳市昊泽空间设计有限公司
设 计 师：韩松
项 目 地 点：广东省东莞市
项 目 面 积：100 m²
主 要 材 料：镜面、石材、壁纸、实木

Design Company: Shenzhen Haoze Space Design Co., Ltd.
Designer: Han Song
Project Location: Dongguan in Guangdong Province
Project Area: 100 m²
Major Materials: mirror surface, stone, wallpaper, solid wood

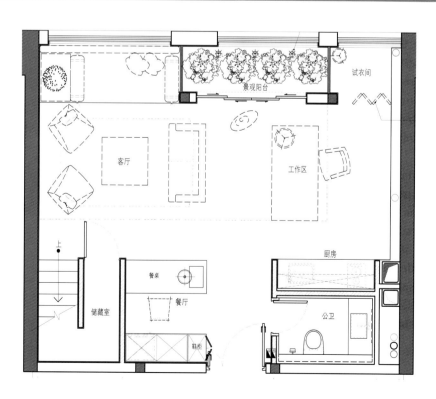

Custom Office Space

近些年随着市场经济的发展，出现了越来越多的自由职业者。他们为了生活辛勤劳动，常常因为工作而降低了生活的品质。本案业主为单身女性婚纱设计师。本案设计师在业主工作需求与生活品质间进行平衡与探索，将本案定位为具有休闲、知性气质的办公场所设计。

二层为女主人的私人空间，主要功能区包含卧室、卫生间及书吧等。在忙碌了一天后，设计师可以在属于自己的天地里，靠在椅子上，品一杯香茗，读一段小说，听一首音乐，工作与生活的结合也可以做到如此曼妙。

本案具有层高上的优势，设计的前期通过平面规划和空间处理，使其成为上、下两层开放式的 LOFT 居室。一层为工作区，同时包含会客接待、Party、展示、厨卫等功能。色彩上考虑到主人的性别及职业特性，采用了以白色为基底并点缀红色的方案。家具灯饰及艺术挂画都经过精心挑选，整体上显得非常高贵、雅致。

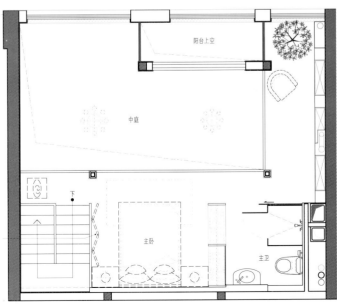

In recent years, with the development of market economy, there are more and more freelancers. They usually work very hard for life and would lower life quality for work. The property owner for this project is a single female wedding dress designer. The designer acquires balance and proceeds with exploration between the property owner's work requirements and life quality, positioning it to be office space design of leisurely and scholarly temperament.

This project has the advantage of floor height. During the former stage of the design, through plane planning and space treatment, the space becomes a LOFT residence with upper and lower two-floor open style space. The first floor is the work zone, with functions such as guests reception, party, presentation, kitchen and washroom, etc. The colors take into considerations the gender of the owner and the professional features, with white as the base, dotted with red graphics. The family lighting accessories and artistic paintings are all through meticulous selection, appearing quite noble and elegant.

The second floor is the owner's private space, with functional areas of bedroom, washroom, study, etc. After a day's work, the designer can retreat back to her own world, leaning on the chair, tasting a cup of tea, reading a novel, listening to some music. The integration of work and life can be so miraculous.

黑白的奏章 · 力宝天马大厦
Musical Chapter of Black and White · Lippo Tianma Building

设计单位：福州国广装饰工程有限公司
设 计 师：张小明
项目地点：福建省福州市
项目面积：200 m²

Design Company: Fuzhou Guoguang Decoration and Engineering Co., Ltd.
Designer: Zhang Xiaoming
Project Location: Fuzhou in Fujian Province
Project Area: 200 m²

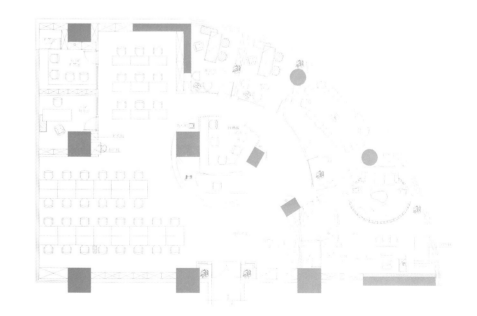

Custom Office Space 313

本案以可持续性发展的理念来构思、规划办公空间，合理地将一些废旧的桌、柜等家具进行改造，然后将其再次利用，践行环保的装修理念。

在空间平面布局上，使动、静空间分配合理，合理地利用空间的采光与通风。在进入有限的入口空间中，点、线、面寻求对称与融合，虚实的黑、白、灰色调在环形的交错空间里显得更加有节奏感和整体感。

玄关区中，吧台成为主导核心，各种设计元素像音符一样融进来，使整体空间层次丰富起来，跳跃而不孤立。黑、白空间就像一首交响乐，在明暗、冷暖、曲直、对立中有节奏地和平共处。

This project carries out conception and planning for the office space with concept of sustainable development, proceeding regeneration towards some abandoned chairs and cabinets in some appropriate way and applying them for a second time, practicing the environmental decoration concept.

In space plane layout, there is the proper allocation of active and quiet spaces, and the suitable application of the space lighting and ventilation. Inside the limited entrance space, the spot, line and surface pursue symmetry and integration and the true and false black, white and gray color tones appear more rhythmic and integral inside the circle space.

For the hallway, the bar counter becomes the core and various design elements integrate inside like musical notes, making the space layers richer, bouncing but not isolated. Black and white space is like some symphony, attaining rhythmic coexistence in dark and bright, cold and warm, curve and straight and contrast.

香港华锋实业 E 路航办公室设计
HongKong Huafeng Industrial Co., Ltd. Office Design

设 计 师：王五平
项目地点：广东省深圳市
项目面积：1600 m²
主要材料：乳胶漆、地毯、玻璃

Designer: Wang Wuping
Project Location: Shenzhen in Guangdong Province
Project Area: 1600 m²
Major Materials: emulsion paint, carpet, glass

本项目位于深圳市南山区科技园德赛科技大厦内，项目面积1600 m²，由三个部分组成。前厅有6m高的挑高空间，原背景是玻璃幕墙，设计师考虑通风以及采光的需求，又要有一面挑高的形象墙，所以设计了五个柱体，与一些横向的线条元素组成了一面大气的形象墙，与白色的logo形象字形成了鲜明的对比，为前台的视觉形象增添了不少感染力。

前台两侧设计了两个接待性的展示区，将地面抬高，与中间部分一起构成了一个"八"字形，也是一个"大路口"的形态。地面做了一个E路航产品的标识，正好指引着"路"的方向，和产品本身的功能寓意一样，使设计的深度与产品的理念相互呼应起来。

大会议有一面形象墙，运用产品的logo元素进行装饰。白色象征简洁及创意感，增强了产品的文化理念。在客户服务区，设计师设计了三个弧形的办公坐位区，吊顶与桌面相互呼应，三个弧形紧连相接，体现了公司工作人员的团结意识。

Custom Office Space

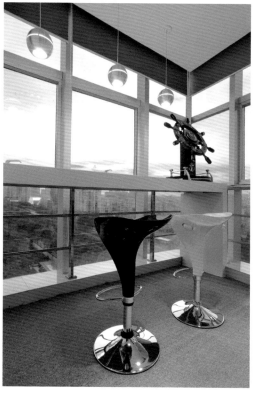

This project is located inside High-tech Industrial Park's Desay Building, in Shenzhen's Nanshan District, with total area of 1600 m² and is composed of three parts. There is a 6m high space in the parlor, with glass curtain wall as the background. Considering the requirements for ventilation, lighting and the high image wall, the designer designs 5 pillars, composing a magnificent image wall together with some horizontal line elements, forming some brisk contrast with the white logo image, while adding some appeals for the visual image of the reception desk.

On both sides of the parlor are two reception presentation areas, constructing some " 八 " shape together with the middle part, also the format of a "grand road junction", with the ground elevated. On the ground is a symbol of Eroda, signaling the orientation of the "road", the same as the functional implications of the product itself, and making the design depth correspond with the concept of the products.

There is an image wall inside the conference room, decorated with the products' logo elements. The white color represents concise and innovative feel, strengthening the cultural concept of the products. Inside the customer service area, the designer designs three arch office seating areas, with the ceiling and the table surface corresponding with each other. The three arches are closely connected with each other, displaying the unity of the whole corporation staff.

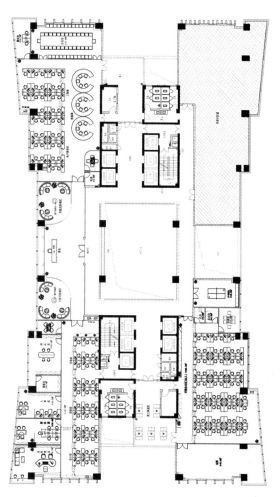

图书在版编目(CIP)数据

定制办公空间/ ID Book图书工作室编 —武汉：华中科技大学出版社，2014.1
ISBN 978-7-5609-9440-6

Ⅰ. ①定… Ⅱ. ①I… Ⅲ. ①办公室－室内装饰设计－中国－图集 Ⅳ. ①TU243-64

中国版本图书馆CIP数据核字(2013)第238202号

定制办公空间

ID Book图书工作室 编

出版发行：华中科技大学出版社（中国·武汉）
地　　址：武汉市武昌珞喻路1037号（邮编：430074）
出 版 人：阮海洪

责任编辑：赵爱华　　　　　　　　　　　　　　　　　　　　　　　　　　　　责任监印：秦　英
责任校对：曾　晟　　　　　　　　　　　　　　　　　　　　　　　　　　　　装帧设计：张　艳

印　　刷：小森印刷（北京）有限公司
开　　本：965 mm×1270 mm　1/16
印　　张：20.25
字　　数：164千字
版　　次：2014年1月第1版第1次印刷
定　　价：338.00元(USD 69.99)

投稿热线：(010)64155588-8000 hzjztg@163.com
本书若有印装质量问题，请向出版社营销中心调换
全国免费服务热线：400-6679-118 竭诚为您服务
版权所有　侵权必究